"研究生学术论文写作"丛书

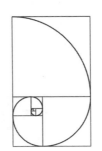

数学科学与技术研究论文写作

案 例 与 方 法

◎主　编　李常品　盛万成　朱佩成

Paper Writing

上海大学出版社

图书在版编目(CIP)数据

数学科学与技术研究论文写作：案例与方法/李常品,盛万成,朱佩成主编.—上海：上海大学出版社，2022.1
（研究生学术论文写作）
ISBN 978-7-5671-4436-1

Ⅰ.①数… Ⅱ.①李… ②盛… ③朱… Ⅲ.①数学-论文-写作 Ⅳ.①01

中国版本图书馆 CIP 数据核字(2021)第 272124 号

责任编辑 丁 译 潘春枝
封面设计 缪炎栩
技术编辑 金 鑫 钱宇坤

数学科学与技术研究论文写作：案例与方法
李常品 盛万成 朱佩成 主编
上海大学出版社出版发行
（上海市上大路 99 号 邮政编码 200444）
（http://www.shupress.cn 发行热线 021-66135112）
出版人 戴骏豪
*
南京展望文化发展有限公司排版
上海普顺印刷包装有限公司印刷 各地新华书店经销
开本 710 mm×1000 mm 1/16 印张 9 字数 137 千
2022 年 1 月第 1 版 2022 年 1 月第 1 次印刷
ISBN 978-7-5671-4436-1/O·71 定价 48.00 元

总　序

　　教育部办公厅《关于进一步规范和加强研究生培养管理的通知》明确指出,研究生培养单位要加强学术规范和学术道德教育,把论文写作指导课程作为必修课纳入研究生培养环节。上海大学积极响应,安排各个学院组织开设相关课程并纳入研究生培养环节,取得良好效果。

　　为了进一步提升研究生培养质量,上海大学研究生院和上海大学出版社联合策划了"研究生学术论文写作"丛书,作为研究生学习学术写作的指导用书。本丛书内容涵盖文科、理科、工科、医学、经济、管理等多个学科,邀请各学科教授及学术骨干领衔担任主编,并根据学科特点,采用以下两种编纂模式:一是对已发表的高水平论文进行综合分析,归纳出写作要点;二是在已发表的论文案例基础上,论文原作者解析撰文过程和注意事项。这种"案例+方法"的编纂模式,通过论文作者现身说法的方式,从问题意识、论证方法、创新之处等方面揭示论文的成之道,为研究生提供可参考、可借鉴的学术写作范例。

　　上海大学老校长钱伟长生前指出,研究生培养分为两个阶段,一个是课程学习阶段,另一个是论文写作阶段。钱校长非常重视研究生学术论文写作能力的培养,他曾经在研究生开学典礼的讲话中指出:"论文很重要。写论文以前,你首先要到第一线找到人家的'肩膀'在哪儿。"本丛书的编纂,践行钱伟长教育思想,探索案例和方法相结合的教学途径,为研究生提供学术研究的"肩膀",为各学科研究生提供学术论文写作的方法指导,也可为青年教师撰写学术论文提供思路启发。

　　我们真诚地希望使用本丛书的教师、学生以及广大读者对其中存在的问题提出修改意见或建议,交流互鉴,共彰学术。

"研究生学术论文写作"丛书编委会

2021 年 9 月

目录

前言

为了积极响应《教育部办公厅关于进一步规范和加强研究生培养管理的通知》，加强学术规范和学术道德教育，把论文写作课纳入研究生的培养环节，2020年9月上海大学研究生院与上海大学出版社召开研究生论文写作教材建设交流会，李常品作为理学院代表，有幸受邀参加，根据交流会精神，负责主编数学学科研究生论文写作课教材。

经过充分讨论和酝酿，该教材由李常品、盛万成、朱佩成联袂主编，书名定为《数学科学与技术研究论文写作：案例与方法》。三位同志多次商议，确定并邀请有丰富写作经验的海内外知名专家学者撰写文章。他们是欧洲科学院院士、IEEE终身会士、香港城市大学讲座教授陈关荣博士，美国南密西西比大学终身教授丁玖博士，国家高层次人才、美国北卡罗来纳州立大学终身教授景乃桓博士，浙江师范大学韩茂安教授，吉林大学马富明教授，西安交通大学何银年教授，兰州大学伍渝江教授，北京应用物理与计算数学研究所、北京大学应用物理与技术中心李杰权教授，上海大学姚锋平教授、高楠教授、余长君副教授、Olga Y. Kushel副教授等十二人。

每位专家都非常负责，精心取材，力求完善，认真地总结写作经验。十二篇文章中，要么是对整个写作过程高屋建瓴的总结，气势恢宏磅礴，要么注重

1

写作的各个环节，细致入微，或者兼而有之；有的介绍如何根据评审意见修改论文以及如何回复修改意见；有的回顾整个研究历程，并将自己的研究经历融入其中，故事扣人心弦；不一而足。

在完成本书的过程中，受到十二位作者的倾力相助，得到了上海大学研究生院和上海大学出版社的支持和帮助，特表衷心感谢。同时也要感谢蔡敏老师，陈强、潘春枝和丁译三位编辑为此付出的辛勤劳动。

李常品　盛万成　朱佩成

2021 年 9 月 15 日于上海大学

漫谈科技论文写作

陈关荣（香港城市大学）

《中国青年报》2018年11月1日登载了一篇调查报告[1]，说2 002名被访大学生中的大多数都认同需要接受论文写作的训练，并表达了对提高论文写作能力的急切需求。

今天资讯发达，网上可查阅的科技论文写作心得感想之类的文章成百上千，图书馆里关于写作的综合指导性书籍也不可胜数。由此看来，学生们通过自己阅读来提高写作质量和水平的客观条件是完全具备的。因此，本文不打算再在文献堆里增加一份雷同的指引。

为了让读者从不同视角去体会科技论文写作的一些要点，本文基于笔者多年来大量编辑和写作的经验，漫谈几个值得注意的方面，旨在与青年读者们分享。

本文采取轻松愉悦的谈论方式，内容主要涉及科技论文的题目、摘要、前言、主体和文献等几个关键部分，也谈及几个简单但不可或缺的栏目，如作者、关键词和总结部分。由于没有实例的讨论并不具体，意思也表达不清，文中将用一些数学和系统科学的虚拟例子，但都尽量作了简化，希望不会妨碍其他领域的读者阅读。

下面分八个栏目来展开讨论。

一、题目

对于命题作文来说，内容要切合题目。科技文章的写作往往是相反的，即文章的技术内容写好了，再来确定一个恰当的标题。这时，题目要切合

1

内容。

一般来说，文章的题目应该简单、清晰、准确、雅致。

题目要"简单"，就是尽可能没有多余的字和词。那些删去了也无妨的字词就删掉好了。例如，"On the study of …"中的"On"是多余的，可以去掉。此外，"The study of …"和"To study …"都不是好的英文题目，而用动名词"Studying …"领起文章题目在经典英文文献中则比较常见。还有，简单的题目比较短小，而这样的文章引用率却较高[2-3]。这是因为短小的题目容易记住，也方便引用和推荐。不要动不动就写个三四行字的长标题，生怕说不清楚。这样的标题读者看完了也记不住，打算引用时也想不起来，便会因为不能顺手查到而放弃引用。

题目要"清晰"，就是告诉读者这篇文章是关于什么主题的。一般不必把背景、内容和方法写进去。例如，"Studying the stability of a nonlinear system by Lyapunov methods"中的"by Lyapunov methods"可以也应该删掉。所用方法在文章里介绍，如果这是一种传统方法，则无需在标题中强调。另外，"a nonlinear system"虽然简单但并不清晰：什么类型的"nonlinear system"呢？如果改成"a three-link robot-arm system""a memristive system"之类就清楚多了。当然，文章的题目不需要也不可能把内容表达得非常具体，但是适当的内涵仍是必要的。不然，没有内涵的字词写出来和不写也差不多。

题目要"准确"。这其实是拟题的一个基本准则，无须多说。但是，这要求有时候会和"简单"相冲突。例如，"Designing a new algorithm that combines a fastest algorithm and an optimal algorithm"是一个十分精准的题目。可是，它包含了三个"algorithm"，真有必要吗？改为"Designing a fast optimal algorithm"就好多了，尽管它不像原来那样可以反映出新程序的具体来历，但它确实说出了论文的主要结果，是准确的。

题目要写得"雅致"不容易，但题目生动雅致，则较容易引起注意。例如，"Reporting a new chaotic system"改成"A new chaotic system coined"，虽然意思没变，但是笔调高雅多了。"Yet, another chaotic system"也可表达同样的意思，且有新奇感。读者看到有趣的标题，哪怕是出于好奇心也会打开你的文章看看，他不看的话当然就不会引用和推荐你的文章，更谈不上去学习你文章报告的新成果。当然，拟题时不要东施效颦，更不要当"标题党"！*Nature* 新闻在 2015 年 12 月 14 日报道说[4]，有学者对 PubMed 医学文献数据库从 1974 年到 2014 年 40 年间的科研论文进行分析后，发现诸如"新颖"（novel）、"迷人"（amazing）、"创新"（innovative）和"前所未有"（unprecedented）等词汇在论文标题和摘要中出现的频率上升了将近 9 倍。这些用字有自卖自夸之嫌，一般不获好评。

总而言之，文章题目非常重要。建议花点时间构思一个"短好"标题。

二、作者

千万不能未经同意就随便地把别人（例如你的导师）的名字加到你的文章上、单方面去投稿。你或者以为日后可以给朋友（导师）一个"惊喜"，其实你可能会被认为是在利用别人（导师）的名声去谋取私利。所有作者对文章都负有几乎同等的责任，比如文章得奖时大家会一起去领奖，但文章受到批评时就不能互推责任。某一篇文章对于你来说可能是很适合投稿和发表的，但对于另外一个人来说就不见得是那么一回事。比如同一篇文章，你发表的话人家可能会说"一个研究生做出这样的文章真不错"，但你的导师拿去发表的话人家也许会说"一个老教授了还发这种水平的文章"。对不同人的评价标准是不一样的。另外，一篇文章在审稿后，不要随意增加或者减少作者的名字。如果确有需要，必须向编辑部解释清楚。作者的排名一般按贡献，但有时也按习惯，这在和国外朋友合作时须特别注意。

三、摘要

摘要须简要地归纳文章的主要内容，不夸张也不遗漏。摘要一般不加评论，也不描述历史背景、写作动机、文章重要性等。例如，下面这句话就是摘要：

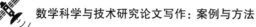

本文证明了哥德巴赫猜想。

但下面这段话就不是摘要，而是一个"扩展摘要"（Extended Abstract）：

哥德巴赫猜想是指任何一个大于 2 的偶数都可写成两个质数之和，俗称"1＋1"问题。该猜想在 1742 年哥德巴赫给欧拉的信中提出，是数论中悬而未决的重大问题之一。过去最好的结果是陈景润 1966 年证明的"1＋2"，但尚不是问题的终结。本文则完全证明了哥德巴赫猜想。

其中很多信息应该是正文的内容，不必也不应在摘要里叙述和描述。很多作者生怕读者不理解该文的意义，在摘要里特别强调所研究的问题如何重要和如何富有挑战性，希望读者特别是编辑和审稿人认可。但这其实不是一篇文章开头写个摘要的目的。一般来说，摘要应该并且只需把文章要点摘录出来，让读者立马知道这是不是他想要阅读的文章，是的话就往下看其内容，不是的话就打住不要浪费时间了。

摘要应该是自我完备的，让读者阅读时不需去查对资料就能理解。摘要出现在文章的开头，写摘要的时候不能假定读者已经看过了文章，因此要避免使用尚未定义的概念和符号。所有内容都须给出说明和描述。摘要中尽可能不要出现数学方程、公式、图表和数据。

文章的摘要非常重要，它向读者展示了文章的全景和主要结果，反映出文章的水平和贡献。建议作者多花点时间拟写一个简洁准确的摘要。

四、关键词

很多人在列出关键词时不加思考，凭感觉随便写几个了事，以为只是应付出版社要求的，无所谓。如果没有必要，为什么科技论文都要求列出关键词呢？关键词应当是反映该文内容最为关键的词，看完了大体上就可以把文章归类存档，相当于 index term。

打个比方，你在图书馆和数据库的"混沌理论"栏目里没找到你本人某篇关于混沌理论的文章，那是因为外行的图书管理员没有从你的文章看到这个关键词。

此外，一个细节是，关键名词只需写单数。作为关键词，"systems"和"equations"等复数名词并不比"system"和"equation"等单数名词提供更多的索引信息。

五、前言和主体

前言，也称引言，是一篇文章最难写好的部分，但是必须写好。

前言一般从所论问题的简要历史背景谈起，引出所要研究的问题，说明该问题的意义、要点和难点，回顾前人的主要贡献，阐述本文解决该问题的基本思路和关键技术，然后归纳出本文的贡献，最后简单勾画全文的结构。

科技文章前言和主体的写作，有几个常见问题需要注意避免：

（1）避免混合使用第一人称主动语态和第三人称被动语态的叙述。两种书写形式的混用会让读者阅读起来有一种换来换去甚至是扭来扭去的感觉，读起来拗口不通顺。例如下面的句子：

We study the Riemann hypothesis in this paper. A preliminary result towards its complete proof is obtained, where we derive a necessary condition. A counterexample is given to show that this condition is not sufficient.

可以改为第一人称主动语态的写法：

We study the Riemann hypothesis in this paper. We obtain a preliminary result towards its complete proof, where we derive a necessary condition. We give a counterexample to show that this condition is not sufficient.

或者改为第三人称被动语态的写法：

The Riemann hypothesis is studied in this paper. A preliminary result is obtained towards its complete proof, where a necessary condition is derived. A counterexample is given to show that this condition is not sufficient.

第一种写法就是你本人一直在说话，让读者一直跟着你向前走。第二种写法是比较文雅和客气的表述，没有"我"和"我们"出现，不强调自己，符合传统英语科技写作的习惯。

（2）避免"抄袭"。不要照抄其他文章（包括自己的文章）中的文句和段落。这在介绍他人的工作时尤其要注意，因为很容易就把人家的原文或原句复制过来。撇开"抄袭"问题不说，一篇文章应该从头到尾用自己的语气写成，不要让人家读起来有一种"剪贴拼凑"的感觉。也不要从自己的旧文中抄录一些语句甚至段落过来。这至少表明，你或者只懂一种句型，或者只会一种表达方式，或者是不认真，或者就是偷懒。其实把一个句子改写一下很容易，何乐而不为呢。

（3）避免写长句子[5]。许多长句的生成是因为加进了插入语，甚至不止一个插入语，诸如 which，that is，with，以及由分词引出的从句。其实写理工科论文的主要目的是传达科技成果的信息而不是让读者去反复品味句子的文采，更应提倡多写短句，让读者一目了然。像下面这样的句子语法是没有问题的，但让人读得很累：

Let us consider a system of three nonlinear ordinary differential equations, where the nonlinearity consists of polynomial functions, which actually are quadratic functions, with only three product terms, that can generate chaos and is well known to be the Lorenz system used to approximate complex weather dynamics, useful for studying fluid motions under various heating conditions.

把它分成两句甚至三句话来说，易读易懂，何乐而不为呢？

（4）避免引进不必要的定义和符号。许多作者过于认真，写科技文章特别是数学文章时，开篇就郑重其事地给出一连串的定义。重要而且在下文反复用到的概念需要预先定义清楚，这无可厚非。但是，一些在下文并不反复使用的概念，尽管可能重要，也不见得需要条列出来。写论文和写教科书是不一样的。教科书需要反复阅读并记忆其中的许多内容，而论文则应方便读者快速阅读。通常读者一看到定义心理就不免紧张，拼命想把它们记住，担心否则文章可能读不下去。可是，一长串的定义和记号会让文章读起来累而无趣，反而妨碍了读者对主题内容的关注和思考。很多文章读完后常常会发现前面好几个定义其实在文中只出现过一两次。这些概念或术语可以在提及的地方再加以说明而不必单独定义。这通常并不碍事，却让文章读起来轻松多了。

（5）避免采用过多的缩写。这在物理类科技文章特别常见。有时看到一篇文章含有十多二十个缩写，诸如 CS、SC、SSC、CSS 等等，混在一起，阅读时需要反复查对前文，十分费劲。采用缩写的原意是让词汇表达简洁从而让文章好读好记，但滥用的结果会适得其反。

（6）避免不必要的方程编号。科技论文中的数学公式或方程式编号仅限于在该文章中使用，是为后面行文中引述前面数学式子的方便而设立。如果一个公式在文章后面无需引用的话，就不必编号。有些文章篇幅不长，却列出一长串的公式编号，一点也不简洁，甚至让人眼花缭乱，实无必要。

（7）避免不必要和不恰当的自夸。写文章不免要说自己的文章比前人的成果有改进有提高，因而有价值有创新。这是自然的，不然为什么写这篇文章呢？但应注意措辞要恰当，不要也不必过分自夸，把评价留给读者去说。如何肯定自己的成果而又能让读者信服，这种好的笔法是要认真斟酌和推敲的。前面提到[4]，文章标题和摘要不宜滥用夸张的字眼，如"创造性"（innovative）、"突破性"（breakthrough）之类，在行文中也不宜滥写过分张扬的句子，如"It must be emphasized that our result is very important"之类。这些词句其实都不是必要的。

（8）避免对别人锋芒凌厉的评判。很多时候为了说明自己写文章的动因，即所谓 motivation，需要指出前人研究成果的不足，这也是很自然的。不过值得注意的是，理工科的学术论文基本上是探讨性而不是终结性的，一般都是关于某个问题的研究过程中的进展性报告。因此，偶然看到一些文章的结果不够完整甚至存在某些谬误并不奇怪。指出其中的问题是应该的甚至是必需的，但目的是把问题说清楚，不是去指责别人，因此措辞可以清淡文雅，让作者和读者都欣然接受。诸如 unfortunately wrong、obviously incorrect 之类用语比较刺眼，建议尽量避免或适当软化。应该认识到，对人家的批评口气越凌厉，并不意味着自己文章水平就越高、报告的科研成果就越有价值。

（9）避免胡乱引用文献。须明白为什么要引用文献，应尽量准确完整地引用必要和重要的参考文献。不要跟着其他文章，别人引用什么你也引用什么，照抄照搬。除了自己熟识的文献之外，如有可能的话把其他引文也找来过目一下，避免不恰当甚至是错误的引用。例如，有些文章可能已有勘误或者已经撤稿，就不要闹笑话去引用了。

还要注意，不要引用自己的但与本文无关的文章以增加引用率。这将会给自我留下一个不良记录。

（10）避免书写中式英语。由于语言和语法的不同，很多中文的表达习惯容易被带到英文科技写作中。例如，"令 F 为可微函数，则它是连续的"是中文一个完整句子，容易被写成

Let F be a differentiable function，then it is continuous.

在英文里，因为这里有两个动词，let 和 is，而且 then 又不是像 and 那样的连词，它应该被分为两个句子：

Let F be a differentiable function. Then it is continuous.

这里，前面那句话怎么没有主语了？是不是应该写为 "We let F be a differential function"？

在科技文章里，由于基本上都是作者自己在说话，并且在说自己的事情，如果都写完整句子的话，就会几乎每句话都用 "We" 或 "I" 来开头，诸如 "We do this" 和 "We do that"，非常累赘。因此，在科技论文写作中，许多一般叙述性的句子都把主语省略了，写成祈使句。例如，下面这类无主语祈使句型是常见的：

Consider the following problem.

Assume that function F is differentiable.

Integrate F from 0 to 1.

其实，生活口语中就有许多无主语祈使句，如 "Thank you" 和 "Let it go" 等，十分简洁又不强调自我。

六、结论

结论就是对文章作个总结，归纳本文的主要成果并发表一些评论。这部

分最好有些学术讨论，指出尚存在的和将来还可以继续思考的一两个重要科研问题，但无需展开研究计划和步骤。

写结论时不要简单地把文章开头的摘要复述一遍，不然就是多余的话，没有新的效果并且浪费文章篇幅。

数学和物理 Letter 类的文章一般都不写结论，这是因为文章较短，看完后印象尚深，无需回顾和总结。

七、致谢

文章需要致谢科研基金是周知的，但不要罗列无关的项目资助。此外，除了感谢认真而又有实质性建议的朋友之外，也应该感谢认真而又有实质性建议的匿名审稿人。这就像路人帮了你的忙后你说声"谢谢"一样，尽管你并不知道那个人的名字。

八、文献

文献一定要统一格式。这并不是可有可无的小事，而是十分要紧的，是科技文章印刷出版的一个基本要求。不要简单地把文献从各种杂志抄过来，然后堆放在一起了事。

翻开你的文稿，编辑和审稿人一眼就能看出来，你写这篇文章到底认真不认真。马马虎虎的作者不屑花时间去整理并统一文献的格式，或者是连这最基本的科技文章写作知识都没有。这种文稿会让编辑和审稿人立马产生一个差评印象，即使还没细看技术内容，就已经形成了"写作马虎""质量不高"的感觉，弄不好就已经产生了退稿的打算。

十分重要的一点是，要使用所投杂志的文献格式。当一个杂志的主编收到你的稿件时发现你是按另一个杂志的文献格式来书写的，他很自然会有一个反应：你的文章好像是被该杂志拒了，现在改投到我们这个杂志来吧？这当然不是不可以，实际上大家都这样处理。但是这位主编会觉得你把文章投给他的刊物并无诚意，很可能就是到处试水。他也可能会想，如果你的稿件真的是被该杂志拒了，我这杂志怎么就比它低一等？这时，因为他认定你的文章曾被别的杂志拒了，他处理时会格外小心，甚至会吩咐编委特别严格把关。这很可能导致你的文章被拒，而你还不知道其实是写作不认真和格式马虎惹的祸。

此外，还有一些细节应加以注意。例如，文献的多少要恰当，不要漏掉重要和必要的文献，也不要罗列与本文关系不大的文章。文献中的每一篇文章都要引用到，否则就不要列入。参考资料常常有多种选择，例如引用一些常用数学公式或著名数学定理的出处就是这样。这时应尽量引用容易找到的文献，以方便读者查阅。

总而言之，写好一篇科技论文绝对不是一件容易的事。

写文章固然有许多值得注意的写作技巧，但写作态度更为重要。写文章需要认真，舍得花时间思考、推敲、修改和润饰，不能敷衍了事。不妨想象一下，一位审稿人拿起你的稿件，一看是这么杂乱无章的，他就会揣测：你这作者花一天时间写成这篇潦草稿件，要让我去花一周时间帮你认真审查？他会想，我只是个匿名志愿者，文章发表后是你的，我一点功劳都没有，却要我付出巨大努力，这对我公平么？因此，他很可能也会学你那样，省点时间，随便挑些毛病拒稿了事。相反，如果审稿人看到你的文章写得很用心，格式整齐，笔误极少，立马就会有个好印象。万一中间有些技术问题他拿不准，这时他会相信你。他会想，这个作者写得那么小心，相信这里会是对的，于是他就会让你通过。否则，他会倾向于怀疑你是错的，因为你前面各种格式、文法、符号、标记等等错误太多，他对你已经失去了信心。可是，这技术问题看不清吧又不好说你错了。于是乎，为了保险起见，他就随便在其他地方挑几个明显错误然后把文章拒了。文章"无缘无故"地被拒了，你也许会觉得不公平，因为你文章的技术成果可能确实是有价值的。这种情况，怪他，还是怪你自己？

其实，文如其人。一篇文章往往就能反映出你的总体学术水平和文化修养。杜甫的诗句"文章千古事，得失寸心知"值得铭记。文章一发表，就成为历史，进了文献档案库，再没有加工修改的机会了，所以说是"千古事"。

对其"得失"，你自己心中要明白，不能掉以轻心，免得将来文章发表后连自己都不满意甚至招来一片骂声时后悔莫及。

最后，笔者提倡大家多写短文。

很多高水平的科技论文都很短，即所谓"short but good"。例如，《自然》刊登过多篇后来获得诺贝尔奖的原创论文都是简要的成果公告式的短篇，包括居里夫人的女儿和女婿 1935 年获诺贝尔化学奖的关键文章，只有 620 个单词和一条化学方程式[6]；Watson 和 Crick 1962 年因 DNA 双螺旋结构获诺贝尔医学奖的关键文章，只有 1 100 个单词和一幅图[7]。类似的例子很多，比如纳什获 1994 年诺贝尔经济学奖的"博弈均衡理论"根植于他 1950 年完成的 26 页普林斯顿博士论文[8]，其主要内容于同年在美国科学院院报（PNAS）上发表时只有 28 行文字[9]。事实上，还有更短的博士论文和奠基性学术论文。这些都表明，一篇文章的质量并不是由它的长度来决定的。

短文的一个好处是易读易记，也易作推介。文章是写给别人看的。如果科技文章能写得像散文故事一样引人入胜，让大家喜欢阅读，那么读者自然乐于学习、推荐和引用。这样，你写文章的目的也就达到了。一篇文章可长可短的话，建议写成短篇。

秉承同样的原则，本文也就此住笔。所说皆一家之言，经验之谈，仅供读者参考。

参考文献

［1］王品芝，顾凌文. 大学生不会写论文，怎么破：88.0%受访者认为大学应开设论文写作课程［N］. 中国青年报，2018-11-01(07). http：//zqb. cyol. com/html/2018-11/01/nw. D110000zgqnb_20181101_2-07.htm

［2］Deng B. Papers with shorter titles get more citations. Intriguing correlation mined from 140,000 papers［J］. Nature，2015. https：//www. nature. com/news/papers-with-shorter-titles-get-more-citations-1. 18246

［3］Letchford A，Moat H S，Preis T. The advantage of short paper titles［J］.

Royal Society Open Science，2(8)：150266. https：//royalsocietypublishing. org/doi/10.1098/rsos.150266

[4] Ball P. 'Novel，amazing，innovative'：positive words on the rise in science papers. Analysis suggests an increasing tendency to exaggerate and polarize results [J]. Nature, 2015. https://www. nature. com/ news/novel-amazing-innovative-positive-words-on-the-rise-in-science-papers-1.19024

[5] Moore A. The long sentence：A disservice to science in the Internet age [J]. Bioessays, 2011,33：893.

[6] Joliot F，Curie I. Artificial production of a new kind of radioelement [J]. Nature, 1934,133：201 – 202.

[7] Watson J D, Crick F. Molecular structure of nucleic acids：a structure for deoxyribose nucleic acid [J]. Nature, 1953，171：737-738.

[8] Nash J. Non-cooperative Games [D]. Ph. D. Dissertation, Princeton University, 1950.

[9] Nash J. Equilibrium points in n-person games [J]. Proceedings of the National Academy of Sciences, 1950, 36：48-49.

数学英文写作的注意之处

丁玖（南密西西比大学）

对于从事数学研究的大学教师，写作并发表学术论文和教学一样都是不可或缺的日常事务；对于数学各个分支领域的研究生，写作也是学术训练的一个重要方面。目前，大部分中国学者是用英文写作数学文章的，导致中国早已成为全世界仅次于美国的 SCI 论文大国。但是许多人的英文写作是有不少问题的。在这篇文章里，我将就英文数学写作谈谈我的看法，并对一些需要注意的地方举例说明。

首先我指出，英文写作与中文写作的基本原理是一样的，即：要想说好某件事，一定要有某事说，这被美国已故的数学写作与演讲大家哈尔莫斯（Paul Halmos）列为"写作第一原则"：In order to say something well you must have something to say. 他心目中的"写作第二原则"是 When you decide to write something, ask yourself who it is that you want to reach（当你决定写作时，问问自己预期读者是谁）。每一个写作者，包括数学家们，都要牢记这两项基本原则。

对于母语和外语而言，提高写作水准的有效方法也是一样的。一个放之四海而皆准的妙法很简单，就是"熟读唐诗三百首"。大量阅读是提高写作能力的一大法宝。如果一个人母语的写作能力欠缺，即便能背得出一万个外语单词，他用那种语言大概也写不出拿得出手的文章。所以那些想在数学英文写作中找窍门走捷径的人，在对那些"博览群书"者写下的优美文章前"临渊羡鱼"之后，不要忘了回到家里"退而结网"。

数学写作属于科学写作。我先来概括一下科学写作的一般要素。科学写作的目的是交流学术思想和科学发现，所以在动笔前作者首先要清楚，这篇文章是写给谁看的。要充分收集与本次写作有关的材料以及关键的文献，要仔细斟酌所表达内容的先后次序和逻辑结构。在写作过程中要精确、清晰、简洁、客观地表达想说的东西。初稿完成后，要反复推敲，不断修改，这对初出茅庐的新手尤其重要。为了不断提高写作水平，要养成阅读国外英文学术书刊的好习惯。此外，要熟记本学科以及相邻学科和相关领域的基本术语和常用专业词汇。最后我想强调的是，写文章贵在创新，切忌表达千篇一律。"抄现成的东西"是我们应该终生"弃之如敝屣"的不良作风。

现在我简单谈论文章的结构。一篇数学文章一般由以下几个部分组成：题目、摘要、引言、主体、结论、参考文献。有的文章还有附录。下面对每一部分大致说明一下。

标题（**title**）是用最少的文字表达文章的主要内容，它应该简洁、具体，并且要直接明了。一个佳例是我的博士导师李天岩教授三十岁时与他的博士论文导师约克（James Yorke）教授于 1975 年 12 月在《美国数学月刊》（*American Mathematical Monthly*）上发表的论文"Period Three Implies Chaos"。虽然文章的定理给出的是在比"周期三点存在"更加一般的假设下，逐次迭代一个连续函数的惊人结论，但"周期三点存在"这个特殊假设使得只有四个单词的标题成为可能，而且该题目概括了文章的主要贡献——此文在数学上首次引进"混沌"的数学术语。笔者参与写作的关于一类新分形的一篇文章，标题就是直截了当的分形名字：Sierpinski-Pedal Triangles，读者看了一目了然。另外一些科学名篇的简洁标题包括"混沌之父"洛伦兹（Edward Lorenz）的"Deterministic Nonperiodic Flow"。

由于文章标题特别重要，它需要作者费点心思推敲确定。香港城市大学电子工程系的陈关荣教授在一次关于科技写作的演讲中，举了自己与弟子合作的一篇短论文作为例子，谈论怎样修改标题。这篇后来被广泛引用的文章，初稿完成后的标题是"A New Chaotic Attractor Connecting the Lorenz Attractor and the Chen Attractor"。在文章修改过程中，作者仔细斟酌了标题，发现它虽然表达清楚，但有不简洁的缺点，例如单词"attractor"出现了三次。最终，他们将它改成了既精炼又醒目的短标题"A New Chaotic Attractor Coined"。

文章标题最好不要太笼统，而要具体，少用那些范围太广的学术名词，切忌写成像个书名，如"Numerical Analysis for Fluid Dynamics"。在某些场合，疑问式的标题容易抓住读者的眼球，如"分形之父"曼德博（Benoit Mandelbrot）的开创性论文"How Long is the Coast of Britain?"总而言之，文章标题将给读者留下关于该文的第一印象，如果它不是那么引人入胜的话，潜在的读者可能不会继续浏览文章的摘要进而阅读全文。

摘要（abstract）是关于文章主要结果的一段或几段文字概括。它是读者的领路人，基本要求是短而精，并能提供文章的关键信息。为了让更多的人读之，应该写得让作者所在专业小领域的"母体"领域大同行也能不费力气地看懂。技术上，它应该避免使用数学符号和公式，最好从头到尾用叙述性文字，让读者觉得就像在林中散步时听老朋友聊天一样。摘要通常写在文章初稿完成后，这时，不应该照抄文章引言或结论两节中的有关句子。要重新构思，尽管表达与文章有关部分同样的意思，也要使用不同的词语表达，免得读者有"似曾相识"之感。另外，可有可无的词和句子尽量不用或少用，如"In this paper"。

如果必须在摘要里提到某篇文章，不要像在正文里那样只列出它在文末"参考文献"内的序号，而要详细给出该文所在的杂志名称和发表信息，如发表年份、杂志卷数和期号。这是因为许多报告过的文章摘要也被放进某次会议的摘要集，而摘要的读者读不到文章后面的参考文献。

纯粹数学文章的摘要一般短于应用数学和计算数学，篇幅比较短的文章摘要甚至一句话就能全面概括文章贡献，如 Lasota 和 Yorke 1973 年发表的一篇现代遍历理论经典论文的摘要是"A class of piecewise continuous, piecewise C^1 transformations on the interval $[0,1]$ is shown to have absolutely continuous invariant measures."

好的摘要能吊起读者的胃口，马上就想读文章的第一部分——引言，因此将它写好对迅速传播自己的研究成果十分重要。

引言（introduction）可被看成是扩充版的摘要，它将给出文章的背景、有关的前人研究成果、本文需要解决的问题、求解的思路等。在叙述别人的工作时，既要客观报道，也不应随意贬低，锋芒毕露；当介绍自己的成果时，要用新闻体式的平实语言，概括得体。整个引言要写得一气呵成，开始部分要用浅显易懂的语言引导读者慢慢登堂入室，信心满满地理解整篇文章要解决的问题和采用的思想及方法。要力求避免一开始就罗列一大堆数学符号，因为这容易导致读者对后面的部分望而生畏而止步于此。引言只需提及与本文密切相关的背景和应用，不必大张旗鼓地谈论应用前景的方方面面，免得离题太远。要知道，研究型文章写的是自己的研究结果，不是有关应用历史或前景的综述场所，所以次要甚至无需的材料不要写得喧宾夺主。

引言的中间应该概括文章的主要贡献、好的新结果、论证的方法，以及期待的效果。它的内容包括：提出问题，总结过去，分析困难，解决措施，给出目的等。

引言的最后部分一般是对文章的结构做出说明，给出文章主体部分各小节的基本内容。但是，这里不应该将各节的小标题再写上一笔，而是补充交代小节的目的所在。

文章的引言写好后常需修改，而主要修改之处往往导致"瘦身"，因叙述啰嗦以及多此一举是写作中常有的毛病，容易让第一节变得"虚胖"。再版过一本数学写作手册的一位英国皇家学会会士说过："改进引言的一个可行方法是删去第一句或前几句，因其常为无关紧要的泛泛而谈。"他举了一个佳例：Polynomials are widely used as approximating functions in many areas of mathematics and they can be expressed in various bases. We consider here how to choose the basis to minimize the error of evaluation in foating point arithmetic. 这两句话中的第一句是人人知道的简单事实，不必再提。因此他将两句缩成一句，并且是让人思考的疑问句：In which basis should we express a polynomial to minimize the error of evaluation in floating point arithmetic? 因此，我们只要记住"可写可不写的句子就不要写"的这个简单法则，就能去掉引言中的那些"多余的话"。

主体（main body）是文章的主要部分，它由引言和结论一前一后两节之间的所有小节构成。这几节的写法百花齐放，各有各的特色。一般第一节引入所需的新老概念、术语定义以及预备知识或有用命题；对于计算数学和应用数学分支，它也含有需要数值求解的连续方程，包括微分或积分方程，或算子方程。对纯粹数学而言，下面的各节就要叙述作者新发现的定理，通常做法是将定理的证明也一并带上。对于复杂的定理证明，也可以另设一节，专事证明。也有少数篇幅较大的文章将证明放到文章主体的最后甚至更后的附录中。对于应用数学和计算数学，这里是数值方法设计和分析的场所，包括关于算法收敛性的理论证明，加上计算试验的数值列表画图。

如果定理的证明较长，帮助读者理解关键想法的一个好方法是将一个或几个思路的结果写成引理，放在定理的前面。有的文章将所有定理所需的引理统统放在一起，甚至单列一节。但是较好的做法是将服务于哪个定理的引理放在那个定理的前面，便于读者阅读。

文章主体部分的撰写没有套路可循，但基本的原则就是清楚完整地展示作者解决问题的思路和方法，不同学科的写作风格和表达方式可能稍有区别。初学者可以通过阅读写作经验丰富的同行专家已被发表的论文，最好是英文为作者母语刊登在权威杂志上的那些，来学会怎样写。刚开始从事写作时可以小模仿，但绝不能整段整句地抄别人的叙述。主体部分包含许多数学公式，有的因为后面用到而需要标号。记住一个原则：尽量少用标号。后面不需要的公式就不要标号。如果某个公式仅在下一行或几行处提到，也不必标号，可以用像"the above equality"类似的词组指出该公式而不会引起误解。

结论（conclusions） 通常是文章的最后一节，它的功能就是关于文章内容的"总结性发言"，阐述本文的主要贡献，使得读者在读完文章主体后有个全面的回顾。这时作者可以展望未来，就相关的研究提出设想。理想的结论节除了不重复引言中的句子外，还要具备让读者有余音绕梁之感或意犹未尽之慨，会引发他们对未决问题的思考，甚而心中涌起跃跃欲试的冲动。与引言不一样的是，当在结论节总结文章的主要工作时，常常用过去式时态来叙述，例如"The main contribution of the paper is that we proved an existence theorem for the class of piecewise convex transformations of the unit interval."

一些人写到结论节的标题时只写单数，不加表示复数的"s"，这是不太规范的。按照约定俗成的做法，这一节的标题是"Conclusions"，如果不用其他词组如"Concluding Remarks"的话。

参考文献（references） 中堆放的是与文章内容有关并且被正文引用的那些论文和书籍。它看上去可以不费吹灰之力地完成，但实际上是最容易出错的地方。最大的错误可能是格式不统一，即不同参考文献的写法各异。原因可能是作者不分青红皂白地将别人文章中的对应条目全盘照搬，而忘了不同杂志对参考文献的写法格式或有细微差别。其他错误包括作者的姓和名次序不一致、杂志名称的全名和缩写前后混用、属于同一范畴的词组字体不一样、上下不按照第一作者英文姓的首字母次序排列等。参考文献应尽可能地放上原始论文，除非信息缺乏而迫不得已改列间接出处，如一本书。当然教科书中的经典结果无需列出最早的出处，只要给出含有它们的著作即可。还有一个需要注意的事项是：文章中被引用的全部论文或图书必须与参考文献中的所有条目一一对应。作者应该记住，与文章没有关系的文献一条也不应该列。极端的例子属于柯尔莫哥洛夫（Andrey Kolmogorov），他有约15篇论文没有参考文献，因为它们都是真正开创性的工作。另一方面，对文章研究成果出过力的任何文献，作者都应该列出，缺一不可。

文章的第一页除了标题外，还包含了作者姓名等信息。关于多位作者的署名顺序，数学文章国际通用的约定俗成是按照英文姓氏首字母排序。记得我在美国读博士时所写的第一篇数学文章，初稿将合作者、导师李天岩教授的名字放在第一，反映出尊师的传统习惯。他修改文章时，将两人的署名顺序颠倒过来，并告诉我"按字母顺序署名"是数学界的不成文规矩，与作者贡献无关。从此我和他的所有合作论文都是这样按照规矩办事的。我的师爷约克教授的英文姓是Yorke，所以他的合作文章几乎篇篇他的署名在后，但这不影响他的学术贡献和地位。如果数学文章的署名违反了规矩，那可能是那个排在后面的作者贡献微小，尽管如按此人姓氏首字母应该排在前面。当然对于反序的例外情形，也有别的种种不便明说的原因。

许多文章的最后有致谢部分，感谢对文章有贡献但未署名的人和/或研究基金。有的文章在投稿过程中，某个认真负责的审稿人对改进文章的内容或语言有不可忽视的帮助，这时作者最好表示感谢这位匿名审稿者。致谢部分应该写得简洁，比如I would like to thank可以缩短为I thank。

  下面我略谈一下文章写作的基本材料——单词和词组。虽然科技写作不像文学创作那样追求想象力和抒情，但有一点是共同的，就是避免重复用词。这就要求作者拥有比较大的词汇量，在表达相同或相似的意思时可以选取不同的单词来表达。我和汤涛教授合著的工具书《数学之英文写作》中，有一章列出了数学不同学科的基本词汇表，以及表达相同或相反意思的英文单词和词组。为了提高英文写作的能力和用词的灵活性，就要经常从本专业欧美作品中吸取营养。比如说，养成定期浏览美国数学会网站 www. ams. org 中会员通俗杂志 *Notices of the American Mathematical Society* 上的综述文章甚至新闻简报的习惯。我于 1986 年去美国读博士后的那几年，作为系里提名的美国数学会学生会员，每期收到这个内容丰富的杂志后，我都要浏览一番，好的文章细读，而不是像有的人那样将它扔在一边，不闻不问。读这类杂志的好处多多，不仅可以了解其他领域的一些行情，而且从那些优美的写作中学会怎样写数学文章，还可以更进一步从那些数学家（如陈省身教授）的访谈中，了解他们成长的心路历程和有趣故事。我自己的英文写作能力除了我以前打下的中文写作基础，就主要得益于对英文杂志这样的大量阅读。

  国内学者英文写作中的一个普遍问题是关于冠词以及可数名词单数或复数的用法上。一般来说，当第一次写到某个可数名词时，比方说在定义一个术语的时候，我们可将此名词或词组写成前面有不定冠词的单数形式，如 Let V be a vector space 或 For an orthonormal basis，或写成复数形式，如 Generalizations of the above lemma appear in various articles. 但是当所述名词带有限定语的时候，常常需要写成前置定冠词的单数形式，如 Let $B(H)$ be the algebra of all bounded linear operators on a complex Hilbert space H. 一个学科的名称，如 algebra（代数），属于不可数名词，所以我们这样写：from Banach's Lemma in functional analysis，而不写 in the functional analysis。然而有了一个具体对象后，就要用定冠词了，如 The first part of the book is on the functional analysis of Frobenius-Perron operators and the second one is devoted to the numerical analysis of the operator. 当一个名词第一次以复数形式出现时，一般不用定冠词，除非名词前有限定性定语，如

teaching for sciences 及 teaching for the computational sciences. 当一个对象存在但不是唯一时，通常用不定冠词或复数，而当这个对象很快再被提及时，就可以用定冠词了，此时 the 有着"这个（this）"或"那个（that）"的意思。兹举一例：

Suppose a smooth function has a critical point. If the second derivative of the function at the critical point is positive，then it is a relative minimum of the function.

抽象名词因为不可数，当然不能用不定冠词，如果非用不可，就需要某种"量词"放在它的后头。例如，我们不能写 This is a wonderful news，而是要写成 This is wonderful news 或者 This is a piece of wonderful news。

还有几个与定冠词有关的写法：（i）当指一样东西时，例如某个公式，可以有两种途径。一种是用 the，另一种就省去它。因此 from the inequality（5）和 from inequality（5）都是正确的写法。但如果不列出公式标号，光写 from inequality 就不对，因为这里读者不知道指的是哪个公式，这时可根据上下文写成如 from the above inequality. （ii）当写以某人名字命名的命题时，可以用所有格形式或在人名前加上 the，例如 by Li-Yorke's theorem 或 by the Li-Yorke theorem，但不能写 by the Li-Yorke's theorem. （iii）一些形容词当名词使用时，一般加 the，如 From the above 和 the more，the better。单词 next 常用 the，如 the next theorem，但也有不加 the 的时候，如 We come back to this topic next time。

英文数学文章常用以 Let 或 Suppose 等特殊动词开头的祈使句，尤其在命题叙述的开头部分。这时它的结尾应为句号，即点号，所以 Suppose the inequality is satisfied，then the conclusion is valid. 应该改为 Suppose the inequality is satisfied. Then the conclusion is valid. 祈使句也可以换成其他的形式，如 Denote the matrix by M. We find that 可以写成 Denoting the matrix by M，we find that. 前者是两个句子，而后者是带有现在分词结构的一个句子。

连词 and 的用法也值得注意。几个并列的事物或句子之间用逗号隔开，并在最后一个对象前加上 and，这个 and 之前一般加上最后一个逗号或不加，如 If the first function is continuous, the second function is differentiable, and the last function is monotonic 或同样正确的 If the first function is continuous, the second function is differentiable and the last function is monotonic. 自然连接两个对象时,逗号就不需要了,如 See the paper by Zhang and Zhu,而不是 See the paper by Zhang, and Zhu.

一些作者,可能由于没有时间养成习惯阅读好文章,或没有注意到英文中那些拉丁语"小词"的正确写法,容易在词语写作上出错,尤其是那种不易感知的"小错"。无论"大错"还是"小错",在严谨的写作者眼里都是错误。比如说,我在国内学者的文章中见过正确的拉丁文"et al."(意指"等等人")的每一种错误的写法:et al, et. al., et. al. 只有下面的写法是对的:The book by Carson et al. is regarded as the authoritative text on the topic. 列出多于一人时,可以这样写:The article by Jones, Perez, et al. is well-known, but the one by Jones, Lee, et al. has been more widely cited. 注意 etc. 指的是"等等事物",不可与 et al. 混用。

后面带有点号的拉丁语词或英语缩写,当它出现在句子的最后时,一般省去表示句子结束的点号以免重复,如 We refer the reader to the work of Li et al. 对于"i.e."(即)和"e.g."(例如),它们的前后通常加上逗号,如 The claim is thus proved, i.e., the function is the difference of two increasing ones 和 See, e.g., the monograph [4] for more details.

在数学文章里,应该避免将数学符号写在句子的开头,并且在两个数学表达式之间放上至少一个英文词。例如,句子 h is an increasing function 可以改写为 The function h is increasing,句子 From formula (1), $x=1$, $g(s)<0$ for some s 可以写成 From formula (1) we have $x=1$ and $g(s)<0$ for some s 或 We have $x=1$ from formula (1). The inequality $g(s)<0$ is valid for some s.

在简单介绍文章各个部分写作基本要素及其他注意事项后,我想通过举例的形式来对症下药地修改一些写得有缺点的英文。下面的例子有的来自杂志编辑请我审核的投稿,有的来自学术界同行或朋友们撰写的文章初稿。

1. The total methods of accomplishing the task is ... 这里主语 methods

是复数，而谓语 is 却是单数，并且 total 与 methods 搭配不当。改为 The total number of the methods of accomplishing the task is.

2. The product of consecutive m positive integers is divisible by m！这里虽然没有语法问题，但将表示数量的字母 m 写在形容词 consecutive 的后面是不规范的，所以改为 The product of m consecutive integers is divisible by m！

3. For each positive integer k，clearly know that there are $(n-k)^3$ sequences containing exactly k zeros. 这句的毛病是缺乏主语。与汉语可以省去主语不同，英语的主语一般不能少，除非是祈使句。所以该句改为 For each positive integer k，we clearly know that there are $(n-k)^3$ sequences containing exactly k zeros.

4. n boys and n girls sit in a circle，so that no two boys are next to each other，so do the girls. 首先句子以数目字母 n 开始，其次两个 so 既重复又意思不一样，最后少了一个 and。改写为 Suppose n boys and n girls sit in a circle such that no two boys are next to each other and so do the girls.

5. For any subset A of X that has odd elements 这里作者的意思是 A 的元素个数是个奇数，但表达不好。所以改为 For any subset A of X that has an odd number of elements.

6. Summing over $i=0$，1，…，n，there are n^3 selections. 这里 Summing 的主语应该是人，所以改成 Summing over $i=0$，1，…，n，we see that there are n^3 selections.

7. Combining the results of the above two kinds of counting，the identity is proved. 现在分词 Combing 的主语是人，所以后面的被动语态应为主动语态。所以改为 Combining the results of the above two kinds of counting，we have proved the identity.

8. If $a>0$，suppose $b>a^2$，then $(a，b)$ is a solution. 在 "if … then" 带有条件状语的句型中再插入 "suppose" 子句，反而使整个句子复杂化了。可以改为 If $a>0$ and $b>a^2$，then $(a，b)$ is a solution. 也可以写成 Let $a>0$. Suppose $b>a^2$. Then $(a，b)$ is a solution. 或 Let $a>0$ and $b>a^2$.

Then (a, b) is a solution.

9. Let $k = r - x$, it can be seen that k satisfies (1). 这里祈使句后面的逗号应为句号，而且可以少写几个词。可以改为简洁的 Let $k = r - x$. Then k satisfies (1).

10. For another two solutions, see Example 3. 单词 another 后面只能是单数名词，单词 other 后面单数复数都可以用，如 the other person 或 other people。所以改为 For other two solutions, see Example 3.

11. There exist x, x' in X, $x > x'$. 数学表达式之间最好有英文单词将它们隔开。可以改为 There exist x and x' in X with $x > x'$. 或 There exist x and x' in X such that $x > x'$.

12. There is a recurrence relation of the first order linear nonhomogeneous with constant coefficients. 这里 nonhomogeneous 是形容词，应该放在名词词组 recurrence relation 之前，所以改成 There is a first order linear nonhomogeneous recurrence relation with constant coefficients.

13. Let $g(n, k)$ denote the number of k ordered subsets satisfying the empty intersection property of the n-element set S. 在本句里，这些 k 个 subsets 是集合 S 的子集，所以要将第二个 of 及后面的部分移到 satisfying 前面，这样方便读者理解。可以修改为 Let $g(n, k)$ denote the number of k ordered subsets of the n-element set S satisfying the empty intersection property.

14. We construct an n-digit number such that all of the digits 1, 2 and 3 in the n-digit number appear at least once. 在一个句子中，谓语应该尽量靠近主语，所以 appear 最好放到状语 in the n-digit number 前面。可以改成 We construct an n-digit number such that all of the digits 1, 2 and 3 appear at least once in the n-digit number. 如果避免重复写 n-digit number，可用代词 it 代替第二个 n-digit number 而不必后移，所以本句也可以写成 We construct an n-digit number such that all of the digits 1, 2 and 3 in it appear at least once.

15. Let S be a finite set, $|S| = n$, and k be a positive integer. 这里两个

逗号使得阅读起来不太流畅，仅用一个介词 with 就可以去掉它们。可以改为 Let S be a finite set with $|S|=n$ and k be a positive integer.

16. If $|x-y|=d$ holds, then we say x, y have property P. 句子中的等号可以构成条件从句中主语 $|x-y|$ 的谓语，所以去掉作为主语 $|x-y|=d$ 的谓语 holds 后，从句更简洁。此外 x 和 y 之间的逗号可以用 and 代替。正式写作中，宾语从句前要加上 that。可改为 If $|x-y|=d$, then we say that x and y have property P.

17. Directly using the formula in Theorem 1，note c is nonzero, so the inequality is true. 这里现在分词结构和祈使句合用而没用连词，不合语法规则。可以改为 Directly using the formula in Theorem 1 and noting that c is nonzero，we see that the inequality is true. 或 Directly use the formula in Theorem 1 and note that c is nonzero. So the inequality is true.

18. Let k be a positive integer，$k<n$. 逗号不能用于祈使句，除非后面跟着多于一个的并列句子。可以改为 Let k be a positive integer and $k<n$. 或 Let k be a positive integer such that $k<n$.

19. Easily know that the sum in the above equals 10. 这是正确的中文对应表达"易知上述和等于 10"错误影响英文表达的结果。记住，除了祈使句，英文句子必须有主语。可以改成 Easily we know that the sum in the above equals 10. 或 We easily know that the sum in the above equals 10.

20. There are at least am 1 in each of the first r columns. 这里数字 1 是计数的对象，所以数量 am 后面要跟 numbers of。可以改写为 There are at least am numbers of 1 in each of the first r columns.

21. It can be seen that this number is $\leqslant 6$. 除非在完整的不等式中，这里的不等式符号$\leqslant$最好改成英文词组。可以改为 It can be seen that this number is less than or equal to 6.

22. From the known conditions，we can see that such numbers contain 11. 三个词的 we can see 不如两个词的 we see 简洁，而且意思基本不变。可以修改成 From the known conditions，we see that such numbers contain 11.

23. n lines are placed in the plane, where there are no two lines are

parallel. 句子应以单词开头，第一次提到平面时应用不定冠词，从句中出现了两个谓语。改正后的句子是 There are n lines that are placed in a plane, where no two lines are parallel.

24. Then t obviously exactly divides the left side of (2). 两个副词挤在一起，仅仅第二个修饰动词 divides，可以把第一个动词移到 t 之前，改写成 Then obviously t exactly divides the left side of (2).

25. Then any common divisor of a and b is the divisor of their greatest common divisor. 因为一个整数的因子不一定是唯一的，所以不应该用定冠词，而应该用不定冠词。可以改为 Then any common divisor of a and b is a divisor of their greatest common divisor.

26. If both sides of (3) are integrated over the interval, obtain the desired equality, which is the direct deduction of (2). 条件句一般具有固定格式 "if ..., then ..."，谓语前要有主语，deduction 用在这里意思不吻合，应改为可数名词 consequence。可以改为 If both sides of (3) are integrated over the interval, then we obtain the desired equality, which is a direct consequence of (2).

27. From $(a, b) = 1$, we can know that there exist integers m and n such that $am + bn = 1$. 不用多余的词是写作的一个基本原则。这里 we can know 可以删去。改为 From $(a, b) = 1$, there exist integers m and n such that $am + bn = 1$.

28. Let $(a_1, ..., a_n)$ be (finite) nonzero integers. 这 n 个整数不必用括号括起，并且单词 finite 多余，因为已知给出的是有限个数。可以改为 Let $a_1, ..., a_n$ be nonzero integers.

29. In addition, if take $a = 5$, then there is $b = 25$. 条件从句无主语，另外 there is 多余。可以改成 In addition, if we take $a = 5$, then $b = 25$. 其他正确的写法包括 In addition, taking $a = 5$, we obtain $b = 25$. 及 In addition, if taking $a = 5$, then $b = 25$.

30. Note that $0 < k < n$, $n \mid k^{-1}(n-1)!$ is not always true. 这里两个数学表达式紧紧相连不太好看，可以这样分开：Note that $n \mid k^{-1}(n-1)!$ is

not always true for $0 < k < n$. 或者用英文将表达式隔开，写成 Note that $0 < k < n$, and so $n \mid k^{-1}(n-1)!$ is not always true.

31. Therefore, when $m > n$, $f(m) < f(n)$. 与上同理，改成 Therefore, $f(m) < f(n)$ when $m > n$. 或写成 Therefore, if $m > n$, then $f(m) < f(n)$.

32. Let $F_k = 2^k + 1$, $k > 0$. For $m > n$, $(F_m, F_n) = 1$. 同理改为 Let $F_k = 2^k + 1$ with $k > 0$. For $m > n$, we have $(F_m, F_n) = 1$. 或 Let $F_k = 2^k + 1$ for $k > 0$. If $m > n$, then $(F_m, F_n) = 1$. 或 Let $F_k = 2^k + 1$ with $k > 0$. We see that $(F_m, F_n) = 1$ for $m > n$.

33. Then from the above equation, we get $d \mid n$. 可以去掉无特别意思的 we get. 改为 Then $d \mid n$ from the above equation.

34. Prime number is the most component subset of natural number set. 这里第一次提到素数，根据上下文，应用复数。词组 most component 意思不清，可以删去。此外所有自然数的集合是一个确定的对象，应用定冠词。可以改为 Prime numbers constitute a subset of the natural number set.

35. There are only a limited number of nonzero terms in the sum. 这里指的是有限个非零数，应用 finite 或 finitely many。可以改为 There is only a finite number of nonzero terms in the sum. 或 There are only finitely many nonzero terms in the sum.

36. Let a and m be integers greater than 1, $a^m - 1$ is prime, then $a = 2$ and m is prime. 某些东西具有某样性质，此性质可用 such that 开头，这样也将 1 和 $a^m - 1$ 自然隔开。另外，祈使句后面的逗号应为句号。可以改成 Let a and m be integers greater than 1 such that $a^m - 1$ is prime. Then $a = 2$ and m is prime.

37. Therefore, from inductive hypothesis that $f(m, n)$ is an integer for all n, then derive $f(m+1, n)$ is an integer for all n. 词组 inductive hypothesis 是特指的，因此要用定冠词。本句不仅缺乏主语，而且宾语从句没有 that。应修改为 Therefore, from the inductive hypothesis that $f(m, n)$ is an integer for all n, we derive that $f(m+1, n)$ is an integer for all n.

38. Hence, $a > 8c + 1$, and multiply by the above expression, we can get (3). 动词 multiply 应用现在分词形式。单词 can 可以删去。可以改为 Hence, $a > 8c + 1$, and multiplying by the above expression, we get (3).

39. That is, (6) is now obtained, and thus derive (5). 第二分句缺乏主语, 可像第一分句那样也用被动式, 以示一致。第二个逗号也可省去。可以改为 That is, (6) is obtained and (5) is thus derived. 或写成主动语态: That is, we obtain (6) and thus derive (5).

40. For example, $0, 1, ..., m-1$ is a complete system modulo m. 这里, 被列出的 m 个整数作为主语是复数, 如果用括号括起来, 可视为一个整体而作为单数。因此改为 For example, $0, 1, ..., m-1$ form a complete system modulo m. 或 For example, $\{0, 1, ..., m-1\}$ is a complete system modulo m.

41. Let $(a, m) = 1$, then $a^{\varphi(m)} \equiv 1 \pmod{m}$. 可以改为 Let $(a, m) = 1$. Then $a^{\varphi(m)} \equiv 1 \pmod{m}$. 或 If $(a, m) = 1$, then $a^{\varphi(m)} \equiv 1 \pmod{m}$.

42. When d runs over all positive divisors of m, m/d runs over all positive divisors of m. 可以改为 When d runs over all positive divisors of m, the quotient m/d runs over all positive divisors of m.

43. We prove that there always exists $a < b$, then the conclusion is proved. 句子可以缩短, 改为 We prove that $a < b$ is always true, and then the conclusion is proved.

44. In the elementary mathematics, there are three basic methods to deal indeterminate equation. 去掉定冠词, 动词 deal 加介词 with 是固定用法, 可数名词 equation 第一次提, 要用复数或在它的前面加上不定冠词 an。可以改为 In elementary mathematics, there are three basic methods to deal with indeterminate equations. 或 In elementary mathematics, there are three basic methods to deal with an indeterminate equation.

45. Then $2a + 2b + 1$, $-2a + 2b - 1$ are all integers. 只有两个物体时, 不用 all, 而用 both。此外两个数学表达式之间用 and。可以改为 Then $2a + 2b + 1$ and $-2a + 2b - 1$ are both integers.

46. If the integers $y > x$, then $y \geq x+1$. 对于这两个整数 x 和 y，给出的是它们之间的大小关系，而不是它们自己。可以改为 If integers x and y satisfy $y > x$, then $y \geq x+1$.

47. Let a, b, c, d are all positive integers, $ab = cd$. 在祈使句中，只能用动词的原型。此外不要逗号，用单词或词组。可以改为 Let a, b, c, and d be all positive integers such that $ab = cd$. 或 Let a, b, c and d be all positive integers satisfying $ab = cd$.

48. Thus, in $C[x]$, $f(x)$ is divisible by $(x-a)(x-b)$. 应该分开两个数学表达式。可以写成：Thus, $f(x)$ is divisible by $(x-a)(x-b)$ in $C[x]$.

49. The polynomials with rational coefficients $f(x)$, $g(x)$ have nonnegative values. 把 $f(x)$ 和 $g(x)$ 移到 polynomials 后面会表达更清楚。可以改为 The polynomials $f(x)$ and $g(x)$ with rational coefficients have nonnegative values.

50. It follows $f(c) > g(c)$, contradiction. 当用 it follows 时，后面要加 that，另外可数名词 contradiction 前要加 a。改为 It follows that $f(c) > g(c)$, a contradiction. 也可以写成 It follows that $f(c) > g(c)$, which is a contradiction.

关于数学英文写作的注意之处，如上我给出了一些具体意见。通过阅读佳作，动手练习，写作能力会不断提高，交流学术、投稿杂志也会感到越来越自信。总之，数学研究与文章写作是相辅相成的，相得益彰的。

浅谈数学论文写作

景乃桓（北卡罗来纳州立大学）

一、引言

我在中国和美国的大学从事数学教育三十余年，怎样培养研究生写好一篇数学论文，是一件值得思考的事情。主要想讲以下几点：

第一，数学论文是怎么回事？

第二，如何撰写数学论文？

第三，关于数学教育的一点思考。

二、文章乃大事

中国传统文化里，从古到今作文章从来都是大事，例如魏文帝曹丕在《典论·论文》中认为"盖文章，经国之大业，不朽之盛事"。写数学论文和通常的作文章在基本道理方面一致，都是作者和读者之间进行心灵和思想的交流，晓之以理、动之以情地进行述说。晓之以理当以科学逻辑思维返现于文字，结合科学的现状与发展，做到有理有据，这是写数学论文的基本要求。在数学论文里谈动之以情是十分必要的，这也是很多数学论文缺乏的，没有情感注入的文章是无法与读者沟通、引起共鸣的，其枯燥或晦涩甚至可能会导致原本优秀的结论无法继续发展。

为什么要写文章，写数学论文？是为了记载一个理论，一个命题，或者一个概念，这是我们数学研究者的学术研究成果的记录。需要通过文章和读者进行心灵上的沟通，讲好我们的数学故事。为了达到这个目的，要融入情感，运用优美的语言把自己的观点和论述展现给读者，并通过严谨的数学证明和详实可靠的论据进行说明。

学术论文和通常的文学作品当然有所不同，前者更多的是以严谨的方式进行陈述，文字运用上也不如文学作品自由浪漫。然而进行说理时，两者都要讲究逻辑性和艺术性，好的数学论文读起来也同样能使读者着迷。

中国是有着悠久文化和历史的国家。中华文明能够传播于世，就是因为有伟大的思想家、文学家、历史学家记载了我们的文明和历史。我也常常以自己能够运用汉语而感到自豪。历史学家司马迁用汉语写下了不朽的著作《史记》，它不仅是一部记载历史和传承文化的史书，还是写文章的典范，赞之为字字珠玑也不为过，其叙述之扼要、用词之精准、情感之丰沛、人物之传神均值得我们在作文章时学习。我读《史记》之时，常常生出感慨：能和司马迁使用同一种文字写文章是一桩多么幸运的事情！如果我们写作的数学文章也能像太史公那般传神，或者即使只是能够略微展现出一点司马先生的遗风，这种与历史偶像的回响共鸣必将使得我们的数学文章不负汉语之名。

三、数学教育的误区

目前在我们的数学教育上存在一个现象，即并没有教会学生如何写文章。举个例子，通常当我针对某个学术问题，布置学生去研读一篇或几篇相关的数学文章，帮助进行问题的求解或者推导，当学生把研究结果写成论文初稿时，我看到的文章基本在形式上是参照着某篇原文，以雷同的结构或者模式写出来的，完全没有自己的风格，体现不出学生的思考和理解过程。这个通病在现在的大学生中十分普遍，为什么我们的中学大学培养出来的学生不能独立写作，一定要照着一个模板才能写文章？事实上，这就反映了数学教育的一个误区，目前的教育还是传统的填鸭式教育，并没有培养学生的创新能力。对学生的要求也仅限于对书本知识的掌握，这是不足以支撑他们将来开展研究的。同时在基本论文写作技能的训练上也是缺失的，学生不具备能够用自己的语言表达自己思维的能力。

在指导研究生学习的过程中，有时我会布置他们写读书笔记，来测试和提高其写作水平，一般国内的数学教育不会有这个内容。同样是在读书笔记中复述一个定理，美国研究生通常会比中国研究生做得好。这恐怕是美国大学生从进入学校开始就在不断地训练如何写读书笔记，因此其概括能力远超未受训练的中国学生。当然写数学文章如果只是复述已有的理论，不是我们的教育目的，我们需要培养的是能写合格乃至漂亮数学论文的研究生。

四、如何撰写数学论文？

数学论文的核心是数学问题，研究是围绕着问题展开的，因此阐述问题的来龙去脉是不容忽视的。讲好我们的数学故事往往就是从一个引人入胜的问题开始，这也是数学研究的魅力所在。一个好问题，乃至一系列问题，编织了人类抽象思维的网络，对这些问题的研究结果是人类对自然规律的归纳与记录。

既然数学论文是围绕问题展开的，如何把问题描述清楚，把研究问题的过程阐述清楚，也是有原则有方法的。原则其实就是求真，这是每一个数学工作者对自己的基本要求。不论是写数学论文本身，还是在研究数学问题的过程中，只有做到了求真，才能在数学研究的道路上走得更远。

数学论文的形式是多样化的，但是其要素和结构是有要求的。文章的要素主要包括问题、理论、论据、证明和结论，结构上可以划分为引言、正文和尾声。文章的要素是体现在文章结构中的，如果说文章的结构好比人体的骨骼，那么要素就是血肉，有机结合要素与结构，才能获得一篇有血有肉的漂亮文章。下面展开说一下数学文章的结构与要素。

1. 引言

要重视引言部分的写作，引言往往是一篇文章的纲领，主要回答为什么写文章，可以包括提出问题、解决问题的方法以及意义。引言要达到使人读了引言能对你研究的数学问题有一个大致了解的目的，读者这时要决定是否愿意继续读你的文章，编辑也会从这里了解你的文章解决了什么问题，是否应该继续考虑你的论文，是否值得送审。引言要说明为何要写这篇论文。举例来讲，比如可以首先提出研究问题的起源和动机并阐述研究的意义，然后简要介绍论文的主要定理和结论，以及说明和证明使用的方法，而现代数学的发展强调各领域的相关联系，很多结果是用另一个角度去描述一个已知的结果，从中得到启示，然后引出新数学理论的发展。这时引言就十分重要，需要说明已有的结果和新结果之间的联系和异同，要着重说明为何新的观点研究该问题是必要的，这样做将会带来什么可能的新发展。

我曾经和合作者写了一篇关于对称函数的文章，第一次投稿遇到评审返回的评论不好，在仔细阅读评审意见之后，发现评审人并没有明白文章的意义，认为我们只是使用不同的方法再现了一个已知的结果。于是我们修改了引言，针对误解，强调了论文的主要思路是链接了两个十分活跃的领域，并采用类比法说明了用类似的结果如果在另一个领域里将会带来重要进展。然后重新投递后，该文得以顺利发表。

2. 正文

正文是论文的"干货"，是数学工作者的工作体现，在叙述上要做到有理有据，重点突出。正文一般首先简单回顾问题的预备知识，然后按照研究内容依次或者递进地进入主题，这也是研究最主要的工作，这部分要写明主要的论点，并逐一进行清晰论述与证明。

在描述一个数学理论时，行文要简洁，能够删去文字达到同样效果的一定要删去赘文。同时数学文章要注意名词或概念第一次出现时一定要进行说明，让读者能够得知其定义，这是讨论问题的基础。

在介绍数学理论时，通常采取的方法不外乎三种方法。第一种方法是由简单特殊的情况开始铺垫，通过具体实例提出概念，然后由浅入深逐步推进，优点是便于理解，也容易厘清定理的要点。第二种方法是从最基本的定义或公设出发，逐步推导、证明命题，由抽象到具体然后到达目标理论，这和中学学习平面几何的过程类似。第三种方法是混合方法，并穿插图表来解释说明，通常离散数学组合论的文章或者数学物理方面的文章采取这一方法。有人观察到现在发表在理论物理界的超一流刊物 *Physical Review Letters* 上面的实验性论文都少不了精湛的图表。

现在数学专业的研究生的主要学术语言是英文，写作英文论文和中文论文的逻辑与结构是相似的。但是现在的学生图简便，在英文词汇的运用上不能做到生动和多样化。举个例子，在数学论文里最常见的"证明"这个单词，很多学生在文中从头至尾只使用"prove"。其实英文可以有多种说法，比如 prove, show, demonstrate, explain 等，可以穿插使用。当遇到并列的或类似的证明，也要动脑筋尽量使用不同的词汇避免重复一个单词，一个自然段内讲述问题时类似的语句尽量采用同义词，而不是只用一个词。

论点和论据是论文要素里最重要的，如何组织好论文的论点和论据呢？我们写论文基本是从论点出发的，论据是证明论点的。在讲述一个论点时，要抓住中心议题，把最主要的论点或定理确定下来，即中心论点。然后从中心论点出发逐渐展开，通过严谨的论据、论证，来讲述整个内容。有的同学会说，我也想知道我的中心论点是什么，但是我的确不知道自己在写什么。这种情况属于没有重点，还不具备写作条件。写文章一定要突出主题，否则整个论文会变得杂乱无章，读者也无法从你的文章中获取养料。耶鲁大学 Ulrich 认为[1]，批判性思维是写好论文和阐明论点的有效方法。批判性思维具有三个特点：独立思考、反思质疑、开放兼容，可以帮助我们尽快找到中心论点。

其实一个定理的讲法不一定要遵循什么样的定式，只要能够把定理描述清楚，有时加上一个例子就能帮助读者很快理解内容。通常我们也会对定理内涵加以说明，如果定理是推广了前人的结果，那么就需要加以说明，解释一下为什么推广了以前的结论。

一篇论文往往会由数个定理或命题组成，描述这些定理是要分主次的。主要定理是要回答前言中提到的主要问题，其他的定理也许是为了得到主要定理而存在的，它们本身也可以是一个漂亮的结果故而也成为一个定理。这就好比主要定理是文章的顶梁柱，其他定理从逻辑上层次上起到支撑作用，整体形成了该文的大厦。

论证也是论文的要素之一，即在布局和描述整个理论时要假设合理、证明充分。讲述定理的证明时要顾及读者的水平，以什么样的基础展开，从什么理论开始推导都是要思考的问题。一般各专业的论文都有一个约定俗成的规范，论文写作的基础理论一般也以专业中常用的基本材料为基础，加上相关论文的结论都是可以引用的材料。

3. 尾声

有的数学论文还有一个结尾部分，特别是现在数学物理方面的文章通常加上这一结尾部分，总结文章解决了什么问题，在此也可以提出一些未来研究的展望。

4. 其他

文章写好初稿后要仔细修改，好的文章都是经过数次修改而成的。修改时要细致推演，多问一下命题的内涵和外延是否准确，实例验证定理是否正确等。数学文章中的索引包括文中的数学式索引和文末的文献索引，都要细致检查，做到细节不出差错。在论文中难免会提到前人的论点，有时就会出现不同观点甚至冲突，我们在评论他人的结果时态度要客观、语气要平和，做到有理有据，避免出现攻击性，这是对他人工作的尊重，也是数学工作者的基本素养。

综上，数学论文无外乎强调文字简洁，语句通顺，逻辑正确，推导严谨。如何写好数学文章，还是要多读专著，了解大家的写法，学习从字里行间去欣赏。比如 I. G. Macdonald 的 *Hall Polynomials and Symmetric Functions* 是代数组合论方面的经典名著，笔者每一次读都深感其文笔之优美，描述之精炼。又比如另一部经典，R. Stanley 的 *Enumerative Combinatorics*，笔者也从中收获不少。另外必须提到 Macdonald，他写了两篇著名的论文，由此催生了两个蓬勃发展的新数学分支。他具有深远的洞察力同时也是文笔大家，能够把数学分支间的奥妙揭示出来。其一，1972 年他在 *Invent Math* 上发表了一篇论文，阐述了李理论中的分子公式（denominator formula）有一系列重要的 q -级数等式推广形式，这些新等式奇妙地一一对应于无限维仿射根系，并由此发展出了仿射 Weyl 群的概念。正是在此基础上，V. Kac 才得到了著名的 Weyl-Kac 特征标公式，之后无限维李代数成为现代李理论的一个主要研究领域。Macdonald 另一篇经典论文是 1988 年发表在 Strasboug 会议论文集的"一类新型对称多项式"，该文引领了代数组合论三十多年的发展。再如，华罗庚先生写的《高等数学概论》也是一部经典，是为中国科技大学创校编写的教材，其中对微积分的基本内容，他写出了自己的想法。还有一种著作是以内容制胜的。比如 H. Weyl 写的 *Classical Groups*，总结了李群李代数的基本内容，是通过材料的取舍、内容的编排，成为李理论的经典之作。该书教导了几代数学家，从李理论、表示论、不变量理论等多方面展现了李理论的精华。陈省身先生就一直强调读经典，特别是研读大家的原文，这是理解数学理论和学习写数学论文的要素，读原文的好处是能更好地洞悉理论发展的源泉，把握研究的精髓。总之，多读多练习并且多思考，一定可以写出好的数学论文。

参考文献

［1］Emily Ulrich. On Academic writing. Interview（Youtube），2017.

论文选题与写作漫谈

韩茂安（浙江师范大学）

1996 年 10 月，我从山东科技大学（原山东矿业学院）调入上海交通大学，作为硕士生导师，每年招收 2～3 名硕士研究生，1998 年开始招收博士研究生，每年 2～5 名；2005 年 6 月，我调入上海师范大学，每年招收 2～4 名硕士及博士研究生；2018 年 10 月受聘于浙江师范大学。迄今已培养研究生 100 多名。作为导师，我颇感疲惫的工作有两个：一是为研究生拟定研究课题（对大部分研究生都如此），二是修改研究生的英文论文（一般要修改 10 遍左右，有时比自己亲自动手完成一篇论文要耗费更多精力）。这样的做法我已经坚持差不多 30 年。当然，未必对整篇论文全部进行修改，而是有所侧重。具体而言，就是对论文前面部分进行比较详细的修改，中间部分在相应位置做明确标注，最后面部分就相对粗略一些。修改的内容，首先是论文的结构与论证的思路，其次是推证过程中的表述，最后是语法与用词，包括标点符号等等。我给研究生反复修改论文的一个主要目的是手把手教他们如何修改并完成论文。

很多情况下，论文修改时会遇到学生写作方面的问题，而且这些问题反复出现，这是让导师感觉身心疲惫和一定程度上厌倦的主要原因。为了提高研究生的写作能力，2011 年，我撰写了一篇题为《课题研究与论文写作技巧》的指导性文章（该文于 2013 年发表在《数学文化》杂志上，见文[1]），给我的每位研究生发了一份，希望他们认真阅读，提高论文写作能力。然而，效果并不明显。

这促使我进一步思考：如何才能较快地提高研究生的写作能力？然后，我有了一个新想法：写一本关于写作训练、写作实践和写作指导的教材，为他们开设一门写作指导课程。于是我从 2016 年开始动笔，经过近两年的奋战，完成了书稿《数学研究与论文写作指导》，并于 2018 年 7 月由科学出版社出版，见文 [2]。该书的出版得到了出版社责任编辑张中兴老师的大力支持。

从 2018 年开始，我利用业余时间给研究生主讲"数学写作指导"课程，得到了浙江师范大学数学与计算机科学学院的重视和认可，并于 2020 年将"数学写作指导"列入研究生的必修课。

之后，我又与浙江师范大学的两位同事一起录制了 25 个慕课视频，建立了该课程的网络平台。在科学出版社张中兴老师的策划下，在 2021 年初又重印了《数学研究与论文写作指导》的教材，将这 25 个微课视频收入书中，读者可以扫码视听。开设该课程分三个阶段：第一阶段是写作基本训练，主要是老师讲题和分析（引导学生如何从一个例题思考出一系列新问题）、学生做题（学生课前做题、制作 PPT）、课堂交流（安排学生讲题、讨论修改、老师点评）。第二阶段是论文精读与分析，安排学生自学 2~3 篇数学研究论文，分析论文的结构，总结论文的结果与方法，也安排学生报告、老师组织讨论。第三阶段学习论文写作规范、学习知名专家的论文摘要与引言范例和英文常用词、拟定课题任务、完成一篇论文习作（这个论文习作可作为写作指导课程的考试卷）。

下面我根据自己做研究和指导研究生的经历，就数学论文的选题和写作所涉及的几个方面，谈谈自己的体会及应该注意的问题，供广大研究生同学以及青年教师参考。由于本人水平有限，所言所为肯定多有不足，欢迎读者不吝赐教。

一、关于论文选题

数学类硕士研究生学制大多为三年，一般可分为两个阶段，第一个阶段是修课学习打基础，第二个阶段是做课题写论文。第一阶段要学习一系列课程，并修满学分。这些课程又分为三类，一类是数学公共基础课，一类是专业基础课，一类是非数学课，其中专业基础课往往是导师亲自主讲。我长期坚持给常微分方程方向的研究生开设了两门课，一是"常微分方程定性理论"，一是"极限环理论"或"极限环分支理论"。坚持主讲这些课程的一个重要收获是在 1994—2013 年间先后出版了一系列著作，见文 [3-8]。研究生修读这些课的目的是打好基础，为顺利完成学位论文做准备，也为毕业后继续从事数学事业做准备。当然，能否打好基础，有两个关键因素，一是老师是否教得好，一是学生是否学得好。

第一阶段之后，紧接着就是第二阶段，即做课题写论文。然而，在做课题之前需要有半年左右的预备期，即研读文献、举办研讨班、拟定研究课题。我在文 [1-2] 中给出了如何读文献的建议，即要就问题由来、所用方法和创新之处进行归纳总结和深层次思考，认真写出自己的读后感悟，并思考回答下列问题：

（1）本文的主要结果、主要方法和主要创新点各是什么？

（2）本文的方法能否用来解决其他问题？

（3）本文的结果和方法能否衍生出新的问题？

一般来说，在导师指导下精读 3～5 篇内容相关的新论文，通过思考上面三个问题，就可以发现比较合适的且有意义的问题。我在讨论班上听了学生的报告之后，都会主动思考，有时把所读论文的方法与自己的方法相结合，想出新课题，而且详细展示自己的思考过程，使学生也学会如何自己找课题。此外，对研究生而言，主动查找一些新文献，用心研读，甚至多读几遍，并在研读过程中思考上述三个问题，写出书面总结，对提高自学能力和写作水平都大有益处。

研究课题及研究思路确定之后，就要想方设法去完成和解决。在工作开始之前，一定要想明白需要解决什么问题，完成什么任务，认真思考用什么思路、方法和工具去解决，即制订研究方案。当然，在解决问题过程中可能遇到没有预知的困难，也可能遇到难以逾越的障碍。出现这种情况，也不必急于求成，而应该静下心来继续思考，再去查阅相关的文献，从中受到启发。如果遇到无法克服的困难，就试着调整思路，绕过障碍。

在解决问题的过程中，对所有的推导与论证要编页码，要保存好草稿。同时，有什么灵感要及时记录下来（有时候灵感就如梦幻一样，瞬间即逝）。

论文选题的过程，就是一个创新的过程，很多时候研究课题与研究思路几乎相随而显。在这个过程中起关键作用的是勇于质疑、善于思考。要养成质疑和思考的习惯，有了这个习惯，就会有做不完的问题。下面，我列举一些例子，谈谈我在数学研究与教学过程中发现的一些问题。

1. Bogdanov-Takens 分支与相关问题

所谓 Bogdanov-Takens 分支是一类平面自治系统余维 2 的尖点在普适开折下的动力学性态研究，其中较有兴趣的一种现象是始于 Hopf 分支终于同宿分支的极限环的存在范围。研究的主要思路是经尺度变换把局部问题转化为非局部问题，然后利用近哈密顿系统的分析方法来获得极限环的存在区间。在最初的一些文献中（例如文 [9-10]），利用所谓的 Picard-Fuchs 方程来得到闭轨族产生极限环的唯一性，并获得了极限环的存在区间。然而，这些文献中没有研究闭轨族的边界（中心奇点与含一个鞍点的同宿环）产生极限环的问题，而由于两个边界是含有奇点的，Picard-Fuchs 方程的方法不再有效。要想完整解决 Bogdanov-Takens 分支中极限环的唯一性问题，研究闭轨族的边界产生极限环的问题是不可回避的。我们最先在文 [11] 中指出了需要研究闭轨族外边界同宿环产生极限环的唯一性，而且为了解决这一问题，先建立了一般的理论方法，即给出了同宿环附近 Poincaré 回归映射有唯一不动点的条件以及受扰系统在未扰同宿环附近存在同宿环的充要条件，然后应用到 Bogdanov-Takens 分支，证明了同宿分支中极限环的唯一性。文 [12] 等多个文献对高余维的 Bogdanov-Takens 分支进行了进一步的研究，并提出了一个扰动引理。这一类分支问题都可以经过适当的尺度变换，化为下列 C^∞ 的近哈密顿系统

$$\dot{x} = H_y + \varepsilon f(x, y), \quad \dot{y} = -H_x + \varepsilon g(x, y),$$

其中 $\varepsilon > 0$ 是小参数。设存在 $h_0 > 0$，使得当 $0 < h < h_0$ 时方程 $H(x, y) = h$ 定义了一族闭轨 L_h，且当 $h \to 0$ 时 L_h 趋于位于原点的未扰系统的初等中心奇点。利用哈密顿函数 $H(x, y)$ 和上述近哈密顿系统的正向轨线可以建立一个形如

$$P(h, \varepsilon) - h = \varepsilon \Delta(h, \varepsilon)$$

的分支函数，文 [12] 给出的扰动引理说，函数 $\Delta(h,\varepsilon)$ 在区间 $[0,h_0)$ 上是 C^∞ 的，并且可以用它来研究 Hopf 分支问题，但没有给出光滑性的证明。这里需要注意，量 h 不是直角坐标，这个光滑性并非显然，也不正确。我们在 [3，13] 中对这个问题进行了细致的研究，给出了反例。在[3，14]中建立了中心奇点产生极限环的一般理论，主要思路是利用直角坐标建立中心奇点邻域内的分支函数以及这个函数与函数 $\Delta(h,\varepsilon)$ 的关系，利用这个关系可以证明分支函数 $\Delta(h,\varepsilon)$ 在 $h=0$ 关于 $\sqrt{h}$ 为 C^∞ 的，但在 [3，14] 中都没有证明首阶 Melnikov 函数 $M(h)=\Delta(h,0)$ 的无穷次光滑性，这一问题在文 [15] 才得到解决，而 [3，13] 中给出的反例是第二阶 Melnikov 函数的非光滑性。对 Bogdanov-Takens 分支中极限环的唯一性问题的完整的讨论先后在 [5-6] 中给出，然而，在外文文献中一直未有见到完整的论述，而且近几年的外文文献在论述这个问题时一直没有意识到要单独研究闭轨族边界产生极限环的问题。鉴于此，我们在文 [2] 中总结了研究中心奇点与同宿环产生极限环的一般理论与方法，给出了 Bogdanov-Takens 分支中极限环的唯一性问题的完整论述，并举例说明在开区间 $(0,h_0)$ 上首阶 Melnikov 函数 $M(h)$ 的零点个数的唯一性不能保证中心邻域内极限环的唯一性，也不能保证同宿环邻域内极限环的唯一性。在 Bogdanov-Takens 分支问题中，出现了三种不同类型的分支，即 Hopf 分支、闭轨族分支与同宿分支，研究这三类分支问题都需要建立所谓的 Poincaré 回归映射，并研究其不动点存在的条件，但在这三类分支中建立回归映射的方法是不一样的，它们的光滑性也大不一样，因此研究其中一个分支问题的方法与结果不能直接用于另外两种分支问题。然而奇妙的是这三类分支问题出现的条件都可归结为 Melnikov 函数的性质（但异宿分支就不同了，由于多个奇点的出现会导致意外极限环的产生）。

2. 中心焦点判定与相关问题

中心与焦点的判定问题是平面微分系统的经典问题之一，对这个问题有三种方法：一是在奇点的小邻域内建立 Poincaré 回归映射，利用这个映射引入焦点量或李雅普诺夫常数的概念，二是 Poincaré 建立的形式级数方法，三是比较新的规范型方法。后两种方法可以利用编程借助于电脑来快速计算焦点量。这里简单介绍第二个方法，并指出其中隐藏很久的一个问题。

考虑平面系统

$$\begin{cases} \dot{x} = \alpha x + \beta y + P_1(x, \ y), \\ \dot{y} = -\beta x + \alpha y + Q_1(x, \ y), \end{cases} \tag{1.1}$$

其中 $\beta \neq 0$，P_1 与 Q_1 在原点邻域内为无穷次连续可微的，且满足

$$P_1(x, \ y) = O(\mid x, \ y \mid^2), \quad Q_1(x, \ y) = O(\mid x, \ y \mid^2).$$

下面的引理是 Poincaré 建立的。

引理 1.1 设系统（1.1）中的函数 P_1，Q_1 为 C^∞ 的，则任给整数 $N > 1$，存在常数 $L_2, \ldots, L_{N+1}$ 和多项式

$$V(x, \ y) = \sum_{k=2}^{2N+2} V_k(x, \ y),$$

满足

$$V_2(x, \ y) = x^2 + y^2,$$
$$V_k(x, \ y) = \sum_{i+j=k} c_{ij} x^i y^j, \quad 3 \leqslant k \leqslant 2N+2,$$

使得

$$V_x(\beta y + P_1) + V_y(-\beta x + Q_1)$$
$$= \sum_{k=2}^{N+1} L_k (x^2 + y^2)^k + O(\mid x, \ y \mid^{2N+3}).$$

此外，如设

$$P_1 = f_1(x, \ y) + O(\mid x, \ y \mid^{2N+2}),$$
$$Q_1 = g_1(x, \ y) + O(\mid x, \ y \mid^{2N+2}),$$
$$f_1(x, \ y) = \sum_{2 \leqslant i+j \leqslant 2N+1} a_{ij} x^i y^j, \quad g_1(x, \ y) = \sum_{2 \leqslant i+j \leqslant 2N+1} b_{ij} x^i y^j,$$

则对 $1 \leqslant k \leqslant N+1$，量 L_{k+1} 仅依赖于 a_{ij}，$b_{ij} (i+j \leqslant 2k+1)$。

上述引理有多种证明方法，一个目前仍在一些教材中出现的经典的证法

需要用到下面的结论。

引理 1.2 对给定的 $\cos\theta$ 与 $\sin\theta$ 的 k 次齐次式 $h_k(\cos\theta,\sin\theta)$，必有 $\cos\theta$ 与 $\sin\theta$ 的 k 次齐次式 $v_k(\cos\theta,\sin\theta)$ 使得

$$\frac{\mathrm{d}}{\mathrm{d}\theta}v_k(\cos\theta,\sin\theta)=h_k(\cos\theta,\sin\theta)-\bar{h}_k,$$

其中 $\bar{h}_k=\dfrac{1}{2\pi}\displaystyle\int_0^{2\pi}h_k(\cos\theta,\sin\theta)\mathrm{d}\theta$。

以往的证明思路是这样的：用待定系数法，将已知函数 $h_k(\cos\theta,\sin\theta)$ 展开成傅里叶级数，则有

$$h_k(\cos\theta,\sin\theta)=\bar{h}_k+\sum_{j\geqslant 1}\big[a_j\cos(j\theta)+b_j\sin(j\theta)\big],$$

而待定函数 $v_k(\cos\theta,\sin\theta)$ 也可以展开成傅里叶级数，即设

$$v_k(\cos\theta,\sin\theta)=c_0+\sum_{j\geqslant 1}\big[c_j\cos(j\theta)+d_j\sin(j\theta)\big],$$

将上面两式代入 v_k 所满足的微分方程，对比同类项，即可求得 c_j 与 d_j，$j=1,\ldots,k$。于是得证。

上面的证明思路很有可能出自 Poincaré，沿用这一思路的证明出现在国内外的多本著作中。我们来思考一下：所需要的函数 v_k 是否已经找到？换句话说，上面找到的 v_k 的傅里叶级数是不是所要找的 v_k？例如，取 $k=2$，能否把 $\cos\theta+\sin(2\theta)$ 写成 $\cos\theta$ 与 $\sin\theta$ 的二次齐次式？答案是不能！当然，上面的傅里叶级数应该不会发生这种情况，但这是需要证明的。在我们发现这个问题之后，就自然要问：引理 1.2 的结论对吗？如果对，如何证明？因为引理 1.1 有不同的证明，其结论没有问题，因此，引理 1.2 的结论也应该没有问题，只是上面的证明思路需要调整或证明细节需要补充，因此，需要给出引理 1.2 的严格证明。我们在文 [17] 中给出了一个很初等的直接证明（也见文 [6]）。读者不妨取

$$h_k(\cos\theta,\sin\theta)=\cos^l\theta\sin^{k-l}\theta,\quad 0\leqslant l\leqslant k$$

来试试。

我们在文［17］中曾怀疑上面利用傅里叶级数的证明思路是有问题的，最近我们又考虑了这个问题，并发现这个思路还是可行的，但需要补充若干细节，下面就来给出证明的主要步骤。

第一步，对给定的 k 次齐次多项式 h_k，记 $h_k^*(\theta)=h_k(\cos\theta,\sin\theta)$。证明 h_k^* 可展开成下列形式的三角级数

$$h_k^*(\theta)=\sum_{0\leqslant n\leqslant k}\big[a_n\cos(n\theta)+b_n\sin(n\theta)\big],$$

其中 a_n 与 b_n 为与 θ 无关的常数。对 k 用归纳法及利用下列三角函数的积化和差公式

$$\cos\theta\cos(n\theta)=\frac{1}{2}\big[\cos(1-n)\theta+\cos(n+1)\theta\big],$$

$$\cos\theta\sin(n\theta)=\frac{1}{2}\big[\sin(n-1)\theta+\sin(n+1)\theta\big],$$

$$\sin\theta\cos(n\theta)=\frac{1}{2}\big[\sin(1-n)\theta+\sin(n+1)\theta\big],$$

$$\sin\theta\sin(n\theta)=\frac{1}{2}\big[\cos(1-n)\theta-\cos(n+1)\theta\big]$$

即可证明。于是，进一步利用傅里叶展开定理可知 $a_0=\bar{h}_k$，$b_0=0$，以及

$$a_n=\frac{1}{\pi}\int_0^{2\pi}h_k^*(\theta)\cos(n\theta)\mathrm{d}\theta,\quad 1\leqslant n\leqslant k,$$

$$b_n=\frac{1}{\pi}\int_0^{2\pi}h_k^*(\theta)\sin(n\theta)\mathrm{d}\theta,\quad 1\leqslant n\leqslant k.$$

第二步，证明若 k 为奇数（偶数），则对偶数 n（奇数 n）有 $a_n=b_n=0$。事实上，由欧拉公式

$$\mathrm{e}^{i\theta}=\cos\theta+\mathrm{i}\sin\theta,\quad \mathrm{i}=\sqrt{-1},$$

或用归纳法易证，对任意正整数 n，都存在两个 n 次齐次多项式 $g_{1n}(x,y)$ 与 $g_{2n}(x,y)$，使成立

$$\cos(n\theta)=g_{1n}(\cos\theta,\sin\theta),\quad \sin(n\theta)=g_{2n}(\cos\theta,\sin\theta).$$

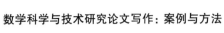

由此可知周期函数 $h_k^*(\theta)\cos(n\theta)$ 与 $h_k^*(\theta)\sin(n\theta)$ 均可写成 $\cos\theta$ 与 $\sin\theta$ 的 $k+n$ 次齐次式，从而当 $k+n$ 为奇数时这两个周期函数在 $[0,2\pi]$ 上的积分等于零，故由 a_n 与 b_n 的计算公式即知结论成立。

第三步。证明 k 次齐次式 $v_k(\cos\theta,\ \sin\theta)\equiv v_k^*(\theta)$ 的存在性。用待定系数法。设

$$v_k^*(\theta)=\sum_{1\leqslant n\leqslant k}\left[c_n\cos(n\theta)+d_n\sin(n\theta)\right],$$

将 h_k^* 与 v_k^* 的傅里叶级数代入引理 1.2 中的微分方程，并比较同类项系数可得

$$nd_n=a_n,\quad -nc_n=b_n,\quad 1\leqslant n\leqslant k。$$

于是
$$v_k^*(\theta)=\sum_{1\leqslant n\leqslant k}\left[\frac{b_n}{-n}\cos(n\theta)+\frac{a_n}{n}\sin(n\theta)\right]。$$

进一步，由第二步的证明知，当 k 与 n 同为奇数或同为偶数时有

$$\cos(n\theta)=g_{1n}(\cos\theta,\ \sin\theta)(\cos^2\theta+\sin^2\theta)^{\frac{k-n}{2}},$$
$$\sin(n\theta)=g_{2n}(\cos\theta,\ \sin\theta)(\cos^2\theta+\sin^2\theta)^{\frac{k-n}{2}},$$

即当 k 与 n 同为奇数或同为偶数时 $\cos(n\theta)$ 与 $\sin(n\theta)$ 均可以写成 $\cos\theta$ 与 $\sin\theta$ 的 k 次齐次式。于是，由第二步的结论即得 v_k^* 的存在性。

在研究焦点与中心判定问题或 Hopf 分支问题时，常常引入极坐标 $x=r\cos\theta$，$y=r\sin\theta$ 把式（1.1）化为一维周期微分方程，再来研究这个周期微分方程的性质，特别是利用周期方程来计算后继函数的展开式系数等。这个周期微分方程有下述形式

$$\frac{\mathrm{d}r}{\mathrm{d}\theta}=R(r,\ \theta),\quad 0<r<r_0,$$

其中 $\lim_{r\to 0}R(r,\ \theta)=0$，而 r_0 是某个适当小的正数。因此，可以补充定义 $R(0,\ \theta)=0$。注意到，如果在极坐标变换中允许向径 r 取负值，则上面的周期微分方程对 $|r|<r_0$ 都有定义了。我们一直都认为如果系统（1.1）中的函数 P_1，Q_1 为 C^∞ 的，那么显然上面的函数 $R(r,\ \theta)$ 对适当小的 $|r|$ 也

是 C^∞ 的。在很多教材或论文中都是这样直接表述的，并没有给出任何解释。易见，上面的结论是下述引理的直接推论：

引理 1.3 如果给定函数 $F(r, \theta)$ 对充分小 $|r|$ 与 $\theta \in [0, 2\pi]$ 为 C^∞ 的，且 $F(0, \theta) = 0$，则存在 $F_1(r, \theta)$，也是 C^∞ 的，使

$$F(r, \theta) = rF_1(r, \theta)。$$

相信很多人都会认为这个引理是显然成立的，不需要证明的。事实上，我以前也是这么想的，根本没想到需要证明它。在一次给研究生上课的时候，突然想到这个结论经常用到，应该证明一下才能够放心使用。就这样，一个小课题就产生了。首先，如果函数 F 为解析函数，则结论是显然的。对 C^∞ 函数，就不那么显然了。对有限光滑情形，结论是这样的：如果 $F \in C^k$，则 $F_1 \in C^{k-1}$。进一步，可以改进泰勒公式的余项形式，例如，如果 $F \in C^k$，$2 \leqslant k \leqslant \infty$，则存在 $F_2 \in C^{k-2}$，使成立

$$F(r, \theta) = F(0, \theta) + r\frac{\partial F}{\partial r}(0, \theta) + r^2 F_2(r, \theta),$$

$$F_2(0, \theta) = \frac{1}{2}\frac{\partial^2 F}{\partial r^2}(0, \theta)。$$

详细证明见 [2, 18]。在研究微分方程的鞍结点分支问题时就需要用到上面改进的二阶泰勒公式，否则，难以说清楚鞍结点分支现象。利用改进的泰勒公式可以证明（证明有点复杂，请读者给出，可参考文 [2]），如果函数 $f : (-1, 1) \to R$ 是一个 C^∞ 的偶函数，则存在 C^∞ 函数 $g : [0, 1) \to R$，使有 $f(x) = g(x^2)$。

关于方程 (1.1)，一个出乎预料但又在情理之中的事实是：如果函数 P_1 与 Q_1 在原点的某邻域内为 C^k 的，$k \geqslant 1$，则所得到的一维周期微分方程中的函数 $R(r, \theta)$ 对充分小的 $|r|$ 也是 C^k 的，证明可在文 [19] 中找到。我们说这一结论出乎预料是因为函数 R 等于一个 C^k 函数和一个 C^{k-1} 函数的比。

3. 平均方法与周期解问题

平均方法是研究一类含小参数的非自治微分方程的动力学性态的重要且有效的方法，其主要思想是利用给定的非自治微分方程，寻求一个适当的自治系

统，通过这个自治系统的一些动力学性态来获得原方程的一些动力学性态，包括周期解的存在性和稳定性、不变流形的存在性和渐近性质等等。为简单起见，这里我们只介绍如何用平均法的思想来研究周期微分方程的周期解问题。

考虑周期微分方程

$$\frac{\mathrm{d}x}{\mathrm{d}t} = \varepsilon F(t, x),\qquad(1.2)$$

其中 $F(t+T, x) = F(t, x)$，$x \in R^n$，$n \geqslant 1$，$T > 0$ 为常数，$\varepsilon > 0$ 为小参数。为叙述方便，假设函数 F 为连续函数，且关于 x 为 C^1 的。引入 F 关于 t 的平均如下：

$$\bar{F}(x) = \frac{1}{T}\int_0^T F(t, x)\mathrm{d}t。$$

这个函数确定一个自治系统如下：

$$\frac{\mathrm{d}x}{\mathrm{d}t} = \varepsilon \bar{F}(x),\qquad(1.3)$$

我们称方程 (1.3) 为 (1.2) 的一阶平均方程。平均法理论的一个主要结论是说，如果平均方程 (1.3) 有一个初等奇点，即存在 $x_0 \in R^n$，使 $\bar{F}(x_0) = 0$，$\det \dfrac{\partial \bar{F}}{\partial x}(x_0) \neq 0$，则当 $\varepsilon > 0$ 充分小时方程 (1.2) 有满足 $x(t, \varepsilon) = x_0 + O(\varepsilon)$ 的唯一周期解。

利用 Poincaré 映射方法很容易证明这个结论。事实上，设周期微分方程 (1.2) 满足初值条件 $x(0) = x_0$ 的解为 $x(t, x_0, \varepsilon)$，令 $P(x_0, \varepsilon) = x(T, x_0, \varepsilon)$，则易求得

$$P(x_0, \varepsilon) - x_0 = \varepsilon T \bar{F}(x_0) + O(\varepsilon^2)。$$

注意到方程 (1.2) 的周期解与 Poincaré 映射 P 关于 x_0 的不动点一一对应，应用隐函数定理即可证明上述结论。另一种证明方法是对方程 (1.2) 引入一个近恒等的周期变换，得到一个新的微分方程，其右端函数关于 ε 的展开式的线性项就是方程 (1.3) 的右端函数。

平均法是研究平面近可积系统极限环个数的重要工具。例如，通过估计平均函数能够出现的单根的个数可以获得相应的平面系统所能够出现的极限环个数之下界。事实上，平均法还可以用来研究一定区域中出现的极限环个数之上界，一些国内外专家也想到了这一点，但是利用的是大家熟知的平均定理，而这个熟知的平均定理根本就没有这样的结论。于是，为了能够利用平均法获得一定区域中极限环个数之上界，就需要发展平均法，建立一个理论依据，即给出用来判定极限环个数上界的方法。于是出于这样的研究需要，近年来平均法理论有了一些新发展，主要有三个方面：一是考虑比方程(1.2)的形式更一般的微分方程，求出了其 Poincaré 映射关于 ε 的展开式

$$P(x_0, \ \varepsilon) - x_0 = \sum_{j \geqslant 1} \varepsilon^j F_j(x_0)$$

中诸函数 F_j 的递推公式。二是利用 F_1 等零点的最大个数来研究方程（1.2）的周期解的最大个数，即第一个不恒等于零的函数 F_j 的零点个数可以用来估计方程（1.2）在一定范围内的周期解的最大个数。三是建立了分段光滑微分方程的平均法理论。我们最近对后两个方面做了一些新的工作，既有一维的周期系统，也有高维的周期系统。详见文 [20-22]。在对一些近可积系统应用平均法研究极限环的个数时，未扰系统一般是等时中心、线性中心或分段线性中心带一个简单的积分因子等，在这种情况下比较容易计算平均函数。为了对更广泛的未扰系统也能够利用平均法，我们还在未扰系统中引入了另一个小参数，这样就有效扩大了平均法的应用范围，见文 [21] 及其参考文献。

以上简单介绍了我们在数学研究与教学中遇到的三个方面的问题和完成的一些相关工作。这些工作是我们在对已有工作认真领会、深入思考的基础上开阔思路、积极探索做出来的。研究生要学习的一个重要功课就是主动查阅文献，然后读懂读透，找出关键所在，经过用心思考就会迸出火花。很多的研究课题和成果都是这样产生的。

二、关于论文写作

华罗庚先生在《数学通报》（1979 年第 1 期）论述了数学研究，提出了

做研究工作的四种境界，即：

(1) 依葫芦画瓢地模仿；

(2) 利用成法解决几个新问题；

(3) 创造方法，解决问题；

(4) 开辟方向。

有很多人的研究始于第一境界，多数大学生的毕业论文也处在第一境界，而研究生的研究工作应该达到第二境界或第三境界，大部分的专家教授应该达到第三或第四境界，当然达到第四境界的人必定是充满活力和具有独创性的。

人们常说，细节决定成败，就是说细节很重要，其实，最重要的不是细节，乃是思想。一旦确立了研究课题，马上就要设定研究目标，即要搞清楚解决什么问题，达到什么目的。据此来设计研究思路，即要用心思考用什么方案解决问题。在问题得到解决之后，就需要把整个过程整理出来，写成一篇学术论文。

学术论文的类型有很多种，各式各样的都有，但常见的数学研究论文一般是涉及理论发展、方法创新、遗留问题、实际应用、进展报告等。论文的写作格式也是多种多样的，并没有固定的样式，但学术界广泛认可的论文写作基本规范也是有的。然而，写作是一种能力，并不是说按照规范去写就能写出一篇好文章的，需要有基本训练和长期积累。例如，我们从大学一年级做数学作业时就应该重视写作，在阅读数学课教材时就应该观察学习文字的运用技巧和数学的逻辑推理。我在文［2］中提出了数学写作的三项基本原则如下：

(1) 结构合理、条理清楚（框架构思）；

(2) 推导无误、论证严密（细节安排）；

(3) 叙述严谨、语句通顺（语言表达）。

这三条基本原则适用于任何内容的数学写作，包括做题、写总结、写论文甚至写书。根据本人长期从事数学教学和研究的经验，对这三条基本原则做如下解读。

第一条：结构合理、条理清楚。证明一个结论，往往有若干步骤，到底分几步完成、每一步的主要任务是什么、每一步出现在哪里等等一定要经过周密思考，并做到心中有数。同样，写作一篇论文往往分若干节内容，到底分几节完成、每一节的主要任务是什么、每一节出现在哪里也要认真思考。有时候某一部分内容可以放在不同的位置，这时一定要想一想安排在何处最合适。

如果是写一本书，就要好好构思一下分几章完成，每一章写什么内容，以及这一章内容分成几节来写。无论是一本书，还是一篇文章，甚至一个章节，都会涉及结构与条理问题，每一部分内容都要力求层次清晰、条理清楚、表述准确、语义连贯、衔接自然。首次出现的记号或概念等都应当及时地给予解释。公式的编排要整齐美观，其中出现的较复杂的表达式可以引入新的记号来代替。段落安排、语句顺序甚至标点符号的使用等都要仔细琢磨。

第二条：推导无误、论证严密。数学推导难免出错，例如，正负号搞反了，系数算错了，有一项给漏掉了，等等。因此，每一步的推导都要反复检查验算，直至确信正确无误。在证明过程中，目标是什么，用什么方法来实现最好有所交代，每一步成立的理由要写清楚。需要时要对公式进行编号，以便后面引用。论证过程要层次分明，要把自己心中明白的东西清清楚楚写出来，使得别人看起来容易接受和理解，尽量不要出现跳跃。

第三条：叙述严谨、语句通顺。即使是数学论文写作，也不能公式一个接一个出现，而应该有足够多的文字的阐述和解释。这样做不但帮助读者理解，增加可读性，还能使得内容华丽有趣，增加美感，使人赏心悦目。因此，所用语言一定要通顺优美，又要通俗易懂且简明扼要。此外，在引用已有的概念和结论时要注明最初的出处，若无法知道最初的出处，至少也要说明在哪里可以找到，以示对原创者和历史的尊重。否则，读者就分不清你引用的概念与结论是别人的还是你自己的。

那么，一般来说，一篇比较完整的论文应该包括哪些内容，每一部分的内容应该怎么去写呢？

首先，论文要有题目（Title）和摘要（Abstract）。题目一般是一个有完整意思的短语，也可以是一个完整的句子，长短要适中，用词要恰当、明了，特别要尽量准确地反映出论文的研究主题（主要结果与目的）。举例如下：

文[15]的题目：On Hopf cyclicity of planar systems.

文[16]的题目：On uniqueness of limit cycles in general Bogdanov-Takens bifurcation.

文[17]的题目：中心与焦点判定定理证明之补充。

文[18]的题目：牛顿–莱布尼兹公式与泰勒公式的拓展与应用。

文[19]的题目：Bifurcation theory for finitely smooth planar autonomous differential systems.

文[20]的题目：Upper estimates for the number of periodic solutions to multi-dimensional systems.

论文的摘要应该是对论文的主要结果与方法的准确的概述，也可以包括对结果、意义等的简要描述，要尽可能简明扼要、意思明确。摘要一般是一段话，少则两三句，多则十来句。摘要又应该以直接客观的文字叙述为主，尽量避免出现复杂的公式以及敏感的注释和评论。例如，

文[16]的摘要：In this paper, we present a complete study of the well-known Bogdanov-Takens bifurcation and give a rigorous proof for the uniqueness of limit cycles.

文[17]的摘要：中心与焦点判定问题的研究是微分方程定性理论课程教学的重要内容之一，然而我们发现在近30年出版的10多种国内外教材中对其中一个主要定理的涉及Poincaré形式级数的证明均有缺陷，本文之目的就是对这一证明做了完整的补充。

文[20]的摘要：The maximal number of zeros of multi-dimensional real analytic maps with small parameter is studied by means of the multi-dimensional generalization of Rouché's theorem. The obtained result is applied to study the maximal number of periodic solutions to multi-dimensional differential systems. An application to a class of three-dimensional autonomous systems is given.

接下来是论文正文的写作，论文正文一般包括引言、预备知识、主要结果与证明等。至于论文正文分成几节，每一节的题目和主题是什么，当然要根据具体情况而定，这需要认真构思，合理安排，并在写作过程中做进一步

的调整。

引言（Introduction）的写作很关键，这部分的内容应该包括以下几个方面：

（1）简洁地提出论文的研究主题、研究动机、研究价值；

（2）介绍有关的历史背景、研究进展和现状；

（3）描述论文所获得的主要结果、所用方法和用途；

（4）简述论文接下来的几节的内容安排，即每一节要研究和解决什么问题。

在介绍研究进展时，有些结果需要给出较详细的介绍，必要时可以把与所做研究密切相关的结果以引理或定理的形式列出，有些结果可以做简单的描述。相关的文献引用尽量全面，语言表述要通顺，充分体现出作者对所研究课题进展和前沿的了解程度和对所涉及的理论方法的理解程度。引言中简述论文获得的主要结果和方法时多用文字叙述，也可以以定理的形式列出（有时候这一节的标题"引言"就可改成"引言与主要结果"）。还可以将论文结果与已有结果进行比较，甚至提出进一步研究的展望，但要注意语言用词，既要委婉优雅，又要实事求是。引言的结尾往往是告诉读者接下来几节的内容组织和安排。引言的篇幅一般是两三页。

预备知识或预备引理（Preliminaries）一般包括两部分的内容，一部分是引用已有的结果或概念，以定义的形式给出或以引理或定理的形式给出，一般不再给出其证明，但应该注明出处，特别是应该注明最初获得这些结果的文献，除非是一些众所周知的经典结果。另一部分内容是建立若干新的命题，一般是以引理的形式写出来，并给出详细证明。

在列出主要结果之前，可以先罗列或证明一些引理，便于后面引用，显得条理清楚，其中个别引理将来还可能在他文中用到。

论文的核心部分是主要结果及其证明，主要结果应该以定理的形式出现，其证明往往是紧随定理之后给出。也有论文的个别定理的证明放在定理之前，也就是说，经过若干步推导之后获得一个你想要的结论，这个结论有独立的使用价值，然后将它以定理的形式表述出来。主要结果可以是一个定理，也可以是多个定理。在叙述定理的证明时，每一步的推理都应该有根据，还应该写清楚证明的思想内涵和结果产生的过程，能够激发和启迪读者的思维。另一方面，数学论文要求语言简洁，表达准确，逻辑严谨，思路清晰。

有时，我们可以在论文正文的最后写一段总结，称为结束语（Summary，Conclusions 或 concluding remarks），再次强调或总结论文的亮点结果或方法，展望未来工作方向。

总之，论文正文每个部分的写作仍是要遵循前面提到的写作三原则。

论文的最后一部分是致谢与参考文献。如果在论文的完成过程中，论文作者受到过一些个人（包括专家、同事、朋友，以及论文的匿名审稿人等）、团体或机构的启发、支持、帮助或资助，那么在论文发表时就应该向他们表示感谢。这就是致谢。"致谢"的英文词是 Acknowledgements（英式）或 Acknowledgments（美式）。例如，文[16]的致谢如下：

Acknowledgments

The authors thank the anonymous referee for his/her valuable suggestions which have helped improve the presentation of the paper. The first and the third authors are partially supported by the National Natural Science Foundation（Nos. 11431008，11771296 and 11571090）and by Shanghai Rising Star Program (No.18QA1403300). The second author is partially supported by the MINECO grants MTM2013-40998-P and MTM2016-77278-P (FEDER) and the AGAUR grant 2014-SGR568.

每个出版社或每个杂志对参考文献的写作格式都有自己的要求，但差别都不大，一般顺序为：作者名、论文题目、杂志名、卷期号、页码等。在编辑参考文献时需要注意两件事：一是所引文献的排序要规范，一般按照姓氏的字母顺序；二是所引文献的格式要统一，避免前后写法不一。

三、论文修改与发表

在完成论文初稿之后，还要进行多次的修改。修改一定要认真细致，逐字逐句进行，连标点符号也不放过。正如文[2]所指出的，在修改过程中，应该注意以下几点：

（1）对多次出现的表达式，或对出现于一个过长的式子里的表达式，应该引入一个新的量（记号）来代替这个表达式。例如，如果表达式 $x^2+2\sin x+e^x+1$ 出现多次，不妨将它记为 $f(x)$，即 $f(x)=x^2+2\sin x+e^x+1$。另一方面，每次引入一个新的量都应该及时地予以解释。对后面需要引用的公式或方程式，应当编号，以便于后面多次引用。如果某公式编了号但文中没有引用，就需要去掉其编号。

（2）每句话、每个公式、每个推理，甚至每段话、每节内容，都要适时给出。例如，某一公式在某处出现，要考虑是不是马上要用到，如果不是马上用到，就应该在后面给出。一些解释性的句子也是一样，应该置于在逻辑上最合适的地方。

（3）每一句话的叙述，特别是定理的叙述，都要语法正确、语意准确、语言精炼、上下连贯、衔接自然，要把自己心中明白的推理清清楚楚写出来，使得读者能够看明白，不至于感觉含义模糊而产生歧义。对于公式推导与证明，一定要条理清楚、逻辑严谨，每一步都应该正确无误、有理有据，并且写明道理，使得读者容易理解，推导与证明过程也要反复验算，以免出错。有些复杂繁琐的计算可以放在文末的附录里。

（4）当完成一个命题的论证或一个主题的表述而转向另一个命题或主题时，需要用到合适的过渡或转折词语来衔接，可以是一个词，一个短语，或一句话。这样可使得上下段的叙述看起来比较连贯和通顺，论证思路比较明确，而不会使读者一下子感到不知所云。

（5）文末参考文献的格式要统一，所列参考文献在文中都应该提及，如果没有引用，就应该删除。

研究生一定要学会自己修改论文，一定要做到逐字逐句去琢磨，从数学与语言两方面都要认真审查每一句话的用词和表述，并自问这样写是否表达清楚了，是否通顺和妥当，是否可以换一个说法，以及如何去改进。

论文写作与论文修改都是很艰巨的工作，写作的时候往往满怀激情，但在修改的时候会很辛苦，有可能感到迷茫而痛苦不堪。论文是写给别人看的，一定要把自己懂的过程写得清清楚楚、明明白白，使得读者能够看懂，而且在阅读过程中得到享受。有时候可能自以为写清楚了，其实可能离写作三项基本原则还相差甚远，因此，论文成稿后需要反反复复的修改。有时候修改几遍之后可以先放一段时间，再看再改，这时候可能有的地方你也感觉不满意或看不懂了，这就说明你没有写清楚，那就再改吧。

我们现在写论文，基本上都用英文完成，除了数学方面的专业问题，还有语言方面的表述问题。以下是研究生用英文写论文经常出现的一些问题。

（1）单复数。谓语有单复数问题，例如，There is a constant. There are two constants. There has a point. There have infinitely many points. 可数名词也有单复数问题，例如 Theorem 与 Theorems，Lemma 与 Lemmas，有些名词的复数形式比较特殊，不是直接加 s，例如，matrix 的复数是 matrices，separatrix 的复数是 separatrices。

（2）冠词的使用。在一个句子中泛指一物时用 a 或 an，特指某物时用 the，例如 We will provide a proof to the theorem. The proof is unique. 也可以说 We will provide a unique proof to the theorem. 又如，We aim to find an upper bound of the maximum number. 有些物是唯一存在的或者是已经确定的，因此其前面的冠词要用 the。例如，the zero vector, the origin, the identity, the integer part of a number, the real part of a complex number，等等。

（3）Let，Suppose 与 Then 的使用。研究生论文中屡次出现的一个问题句型是"Let $a > 0$, then $b > 0$."正确的写法是"Letting $a > 0$, then $b > 0$."或者"Let $a > 0$. Then $b > 0$."有时候可以使用两个并列句，但两句之间应该有连接词，例如，Let $a > 0$ and suppose that the condition of the theorem is satisfied.

（4）时态与语态问题。介绍他人过去完成的工作时要用过去时，介绍本文工作时用现在时。例如，Two years ago he proved the above theorem. In this paper we give a new proof to it. 在一段话中相邻的两个句子应该用一致的语态（都用主动或都用被动）。例如，下面写法是不妥的：Theorem 1 was proved by A, and B proved Theorem 2.

（5）Denote by 的使用。词语 Denote by 是固定搭配，研究生在使用中经常出错，有时忘了写 by，有时没有写对位置。正确的用法如下：Denote by A the average value of the function f over the interval $[0, 1]$. The function has a umique critical point, denoted by B.

（6）从句的使用。在数学英文写作中应该多使用简单句型，少使用带有从句的复杂句型。在宾语从句中连接词最好不要省略不写。例如，Suppose that the function is continuous at $x=0$. 为体现学术论文的严谨，该句中的 that 最好不要省略。

（7）集合的表达。给定正数 a，大于 a 的实数全体应该写成 $A=\{x \mid x>a\}$，而不要写成 $A=\{x \mid x>a, a>0\}$，因为在后一种写法中，可能产生 a 也在变动的错觉。由于 A 依赖于 a，也可以采用下列写法

$$A_a=\{x \mid x>a\}, \quad a>0.$$

（8）误引他人结果。在引用他人结果时叙述别人的某个定理，但却漏写了使结果成立的条件。有时候是因为粗心忽视了条件，有时候是没有看明白别人的结果。引用某一工具性引理时没有注明这个引理是谁获得的，只说可在某文献（不是最初获得该引理的）找到这个引理。这很可能对读者产生误导，也可能会产生二次误导，对原创者也不公平。

（9）懒散。研究生（包括硕士和博士研究生）屡犯本科生不常犯的错误，在论文修改中不认真、不思考，明显的语法不正确、表述不通顺、逻辑不严密、意思不明确、概念不清楚等问题得不到自我纠正。对老师标出的有问题的表达不好好去理解，也不主动问老师（或想问却怕被老师批评），不知道怎么改，于是不改或者乱改。修改论文没有效率，本来一两天就可以完成的，拖了一两个月都完成不了，影响了工作进度，影响了自己的进步、成绩和信誉，甚至会影响个人前途和发展。一些研究生不能按照老师的要求完成布置的任务。例如，老师要求独立完成作业，结果发现雷同。又如，要求在一周内完成一项任务并向老师汇报，结果等了两周也不见人影。部分研究生不能主动查找与研读文献，没有养成主动思考的习惯，做课题缺少激情，工作动力不足。有这些问题的研究生一定要引起高度重视，一定要尽快改正，并振作精神，奋发图强。其实，无论干什么工作，懒散都是成功的大

敌。当你全身心投入学习与工作中，你已经成功了一大半。

修改论文一定不能怕麻烦、不能怕耽误时间，一定要铭记并认真实践"结构合理、条理清楚，推导无误、论证严密，叙述严谨、语句通顺"的写作原则。其实，越舍得花时间反复修改，自己的收益也就越多。在完成修改工作以后，要与合作者商量投什么杂志，并按照杂志的要求投稿。在投稿时要提交一封"投稿信"（cover letter）。下面提供一个例子供读者参考。

Dear Editors,

We are submitting a manuscript entitled "The existence of a limit cycle for a Liénard system" for your consideration for publication in the journal "Journal of Differential Equations". In this paper we confirm a conjecture and obtain a new result as well by using a new technique. We would be grateful if the manuscript could be reviewed.

Thank you for your time to deal with our paper, and look forward to receiving comments from the reviewers.

Sincerely yours,

M. Han (on behalf of the co-authors)

完成投稿后，耐心等待数月就会有杂志编辑部的审稿意见。如果等了四五个月，还没有收到审稿意见，不妨给编辑部写信问问或催催。下面范例可做参考。

Dear Editors,

I am the corresponding author of the manuscript with the number JDEQ19 - 1175. It is more than 4 months since we submitted it for possible publication in your journal. I am writing you to inquire about the status of it. I should greatly appreciate your letting me know the present status of it as soon as possible.

Sincerely yours,

M. Han

等收到编辑部寄回的审稿意见时，恭喜你（们），就有较大可能录用大作了。此时，不能掉以轻心，而应该仔细审阅和琢磨审稿意见，真正搞明白审稿人的意思，然后按照审稿意见做认真的修改，并如期提交修改稿。在提交修改稿时要写一个修改说明（response letter to review reports），其格式如下所述。

Dear Editors and Reviewers,

Thank you very much for reviewing our manuscript with your complimentary comments and suggestions. We have revised the manuscript accordingly. Please find attached a point-by-point response to the reviewers' concerns. We hope that you find our responses satisfactory and that the manuscript is now acceptable for publication.

或者

Dear Editors,

We would like to thank the reviewer for careful and thorough reading of this manuscript and for the thoughtful comments and constructive suggestions, which help to improve the quality of this manuscript. We have studied the comments carefully and made major correction which we hope meet with your approval. We have the following response in details.

在提交修改稿后可能很快就会收到稿件的录用通知，或邀请进一步修改的信函。当收到录用通知时，编辑部可能会再要求你（们）做一些简单的修改，也可能直接录用了，让你（们）等待、关注稿件的排版稿。这样几经周折以后，你（们）的研究成果终于正式发表了。然后，请再接再厉，开始新的创作。

四、基本学术道德与规范

为弘扬科学精神，加强科学道德和学风建设，提高科技工作者创新能力，促进科学技术的繁荣发展，中国科学技术协会根据国家有关法律法规制定《科技工作者科学道德规范（试行）》，并于 2007 年 1 月 16 日在中国科协七届三次常委会议上审议通过。近年来，我国科技发展很快，成就辉煌，举世瞩目。同时，也暴露出一些问题。自 2015 年以来，我国科技界接连遭遇国外出版集团较大规模的集中撤稿，国际声誉受到直接冲击，造成极为恶劣的社会影响。撤稿事件反映出我国少数科技工作者自律意识缺乏，底线意识不强，在名利诱惑面前心态失衡。为进一步加强科技工作者道德行为自律，中国科协在 2017 年又研究制定了《科技工作者道德行为自律规范》。下面简要介绍这两个文件中与科学研究和写作相关的内容。

1. 基本学术规范

进行学术研究应检索相关文献或了解相关研究成果，在论文或报告中引用他人论点时必须尊重知识产权，如实标出。在课题申报、项目设计、数据采集、成果公布、贡献确认等方面，遵守诚实客观原则。对已发表研究成果中出现的错误和失误，应以适当的方式予以公开和承认。诚实严谨地与他人合作。耐心诚恳地对待学术批评和质疑。公开研究成果、统计数据等，必须实事求是、完整准确。合作完成成果，应按照对研究成果的贡献大小的顺序署名（有署名惯例或约定的除外）。署名人应对本人作出贡献的部分负责，发表前应由本人审阅并署名。科研新成果在学术期刊或学术会议上发表前（有合同限制的除外），不应先向媒体或公众发布。不得利用科研活动谋取不正当利益。正确对待科研活动中存在的直接、间接或潜在的利益关系。

2. 主要学术不端

学术不端行为是指，在科学研究和学术活动中的各种造假、抄袭、剽窃和其他违背科学共同体惯例的行为。例如，故意做出错误的陈述，捏造数据或结果，破坏原始数据的完整性，篡改实验记录和图片，在项目申请、成果申报、求职和提职申请中做虚假的陈述，提供虚假获奖证书、论文发表证明、文献引用证明等；侵犯或损害他人著作权，故意省略参考他人出版物，抄袭他人作品，篡改他人作品的内容；未经授权，利用评审机会，将他人未公开的作品或研究计划发表或透露给他人或为己所用；把成就归功于对研究没有贡献的人，将对研究工作做出实质性贡献的人排除在外，无理要求著者或合著者身份。又如，成果发表时一稿多投；采用不正当手段干扰和妨碍他人研究活动，包括故意毁坏或扣压他人研究活动中必需的仪器设备、文献资料，以及其他与科研有关的财物；故意拖延对他人项目或成果的审查、评价时间，或提出无法证明的论断；对竞争项目或结果的审查设置障碍；参与或与他人合谋隐匿学术劣迹，包括参与他人的学术造假，与他人合谋隐藏其不端行为。

3. 科学道德规范

科学道德和学术诚信是科技工作者必备的基本素质，砥砺高尚道德品质是科技工作者的不懈修炼。科研人员要坚持"自觉担当科技报国使命、自觉恪尽创新争先职责、自觉履行造福人民义务、自觉遵守科学道德规范"的高线；又要坚守"反对科研数据成果造假、反对抄袭剽窃科研成果、反对委托代写代发论文、反对庸俗化学术评价"的底线。要牢记并坚持立德为先、立学为本、知行合一、严以自律，严守学术道德和科技伦理，共同营造风清气正的科研学术环境。秉持创新、求实、协作、奉献的科学精神，潜心研究，淡泊名利，经得起挫折、耐得住寂寞，争当学术优异、学风优良、品德优秀的科技先锋。

致谢 杨俊敏、刘媛媛、龚淑华、蔡梅兰、可爱、刘姗姗等阅读本文后提出了一些有益的修改意见，在此向她们表示感谢。特别感谢本书主编李常品教授的邀请，使得我有机会在这里分享我的写作体会。郑重感谢刘喜兰教授认真阅读了该文全稿并亲自做了诸多润色，使该文在叙述上颇有增色。

参考文献

［1］韩茂安. 课题研究与论文写作技巧［J］. 数学文化，2013,4(3)：88-90.

［2］韩茂安. 数学研究与论文写作指导［M］. 北京：科学出版社，2018.

［3］韩茂安，朱德明. 微分方程分支理论［M］. 北京：煤炭工业出版社，1994.

［4］Luo D, Wang X, Zhu D, Han M. Bifurcation Theory and Methods of Dynamical Systems. Advanced Series in Dynamical Systems 15［M］.

Singapore：World Scientific Publishing Co.，Inc.，1997.

［5］韩茂安，顾圣士. 非线性系统的理论和方法［M］. 北京：科学出版社，2001.

［6］赵爱民，李美丽，韩茂安. 微分方程基本理论［M］. 北京：科学出版社，2013.

［7］Han M，Yu P. Normal Forms，Melnikov Functions and Bifurcations of Limit Cycles. Applied Mathematical Sciences 181［M］. Germany：Springer-Verlag，2012.

［8］Han M. Bifurcation Theory of Limit Cycles［M］. Beijing：Science Press，2013.

［9］Carr J. Applications of Centre Manifold Theory［M］. New York/Heidelberg/Berlin：Springer-Verlag，1981.

［10］Chow S N，Hale J. Methods of Bifurcation Theory［M］. New York/Heidelberg/Berlin：Springer-Verlag，1982.

［11］罗定军，韩茂安，朱德明. 奇闭轨分支出极限环的唯一性（I）［J］. 数学学报，1992，3：407-417.

［12］Dumortier F，Roussarie R，Sotomayor J. Generic 3-parameter family of vector fields on the plane，unfolding a singularity with nilpotept linear part. The cusp case of codimension 3［J］. Ergod Th & Dynam Sys，1987，7：375－413.

［13］韩茂安，王政. Melnikov 函数在中心处的可微性问题［J］. 系统科学与数学，1997，17(3)：269-274.

［14］Han M. Bifurcation of limit cycles and the cusp of order n［J］. Acta Mathematica Sinica，1997，13(1)：64-75.

［15］Han M. On Hopf cyclicity of planar systems［J］. J Math Anal Appl，2000，245：404-422.

［16］Han M，Llibre J，Yang J. On uniqueness of limit cycles in general Bogdanov-Takens bifurcation［J］. Internat J Bifur Chaos，2018，28(9)：1850115.

[17] 韩茂安. 中心与焦点判定定理证明之补充 [J]. 大学数学，2011，27 (1)：142-147.

[18] 韩茂安. 牛顿-莱布尼兹公式与泰勒公式的拓展与应用 [J]. 大学数学，2015，31(5)：6-11.

[19] Han M，Sheng L，Zhang X. Bifurcation theory for finitely smooth planar autonomous differential systems [J]. J Differential Equations，2018，264(5)：3596-3618.

[20] Han M，Sun H，Balanov Z. Upper estimates for the number of periodic solutions to multi-dimensional systems [J]. J Differential Equations，2019，266：8281-8293.

[21] Han M，Yang J. The maximum number of zeros of functions with parameters and application to differential equations [J]. Journal of Nonlinear Modeling and Analysis，2021，3(1)：13-34.

[22] Liu S，Han M，Li J. Bifurcation methods of periodic orbits for piecewise smooth systems [J]. J Differential Equations，2021，275：204-233.

在实事求是基础上注重细节
是写好研究论文的必备素质

——研究生如何写好数学论文之我见

马富明（吉林大学）

如果你是一名数学及其相关专业的在读本科生或研究生，并且打算把数学科研与教学作为自己将来的职业生涯，恐怕你就无法躲避撰写数学研究论文这一问题。从大学本科开始，至少你要经历撰写本科毕业论文的过程，如果读研，你还必须撰写研究生学位论文（硕士论文或博士论文两篇之一，或两篇都要写）。在你完成学位后，如果继续从事科研和教学工作，那么你就会不断地撰写科研论文，或指导你的学生写论文。可以说你无法躲避"写论文"这件事的。那么，怎样写论文，怎样才能写出"好"的论文，就成为你迫切希望得到答案的问题。坦率地讲，我没有能力给出这个问题的圆满答案，但作为从教 30 多年，指导过多名研究生的数学教师，我经历过从本科生到教师这个完整过程，同时，多年来为国内外各种学术期刊审稿，为多个高校做研究生学位论文评审，也使我对同学们论文撰写中的不足有所体会。以下是个人的一些看法，我愿意与那些对"怎样写论文"这一问题感到疑惑的同学们分享。

同学们通常面对两种论文写作：专业研究论文——准备投给专业学术期刊发表的，以及专业学位论文——为了获得某一级学位，向学位授予机构（大学或研究所）提交的论文。这两种论文虽然有很多相同的地方，但由于用途不同，还是稍有差别的。

科学研究总的原则是坚持实事求是的科学精神。首先，写论文的初衷是把自己对学术问题的研究成果在"学术共同体"中公布出来，以达到相互交流，促进学术研究的目的。这里的所谓"学术共同体"，指的是对某学科某类问题有着共同兴趣的研究者，这应该是没有异议的，但现实中某些年轻朋友或同学会感觉到并非如此。因为按照这个初衷，论文应该是研究成果的表述，这意味着研究是目的，论文只是科研的一个必要环节，为了科研而写论文。但在当下，在各种利益的驱使下，很多时候的写论文只是为了"发表论文"而写论文，这种写论文目的的异化造成了"垃圾论文"泛滥，也使得同学和年轻朋友们开始写论文时有点不知所措。这个问题的讨论已超出此文讨论的范畴和笔者的能力，我们只好将其搁置。所以，本文下面讨论的只是出于较为单纯目的的论文写作问题。这样的话，"怎样写论文"这个问题的回答似乎就简单一些了。

在坚持科学原则的基础上，论文写作的细节在很大程度上也可能决定你是否能写出一篇合格的学术论文。我认为一篇完整的论文应该告诉读者这样几件事：第一，作者研究了什么问题；第二，作者研究这个问题的动机；第三，作者的研究成果。也可简单概括为三个"什么"，即研究了什么、为什么研究它、得到了什么？下面我们仔细讨论一下这三个"什么"。

这里的第一件事，就是论文要明确地告诉读者，此论文所表述的是作者对哪一个具体数学问题的研究结果。你可以通过合适的论文标题、简短的论文摘要，以及论文正文中准确的数学表述三部分来达到这一目标。这些看起来不那么复杂的事情，对初次撰写论文的同学或许并不那么简单。首先，论文要有一个适当的标题。初次撰写论文的同学易出现的问题是论文标题不合适。例如，标题口气"太大"，像"某某问题研究""某某理论的某某问题研究"。这样的标题通常会使读者不能在浏览论文的标题时获得较为准确的信息，应当避免。举例来说，比如你研究了三维空间一个立方体区域上 Stokes方程的一个有限差分格式的稳定性条件，你的论文标题如果定为《Stokes 方程有限差分方法》可能就不太合适。当然，克服这种倾向的同时，也应避免另一种倾向，就是把论文标题写得太长，太啰嗦。究竟怎样合适，还是得具体事情具体分析，似乎没有所谓的标准。在论文正文中，如何清晰地表达作者所研究的问题（包括已知条件，讨论问题所需的假设，待求的未知量或要达到的其他目标）看起来不应成为问题，但从多年为国内外各种学术期刊审稿的经历来看，也不尽然。特别是研究生同学，在自己投出的论文稿中，不一定能清楚、准确地表达自己研究的数学问题。我认为这里面的一个原因是很多同学平时就不太注重数学表述的准确与严谨，以为自己懂了，别人也应该懂。事实上，由于科研领域的广大，并不是每位读者都能很快抓住你的思路，明白你的问题，如果这位读者恰好是你稿件的审稿人，而你的糟糕表述给了他一个不好的印象，就会降低你的稿件通过评审的可能性，导致稿件被拒。

第二件事，就是应该在论文里表述清楚你研究问题的动机，即为什么要研究这个问题。部分同学也可能感觉到这一点必要性不大。研究成果摆在那，存在即为必要，还要解释吗？我的意见是这点不仅必要，还很重要。通过研究动机的解释，把所讨论的问题的来龙去脉讲清楚，把你认为的问题的重要性讲清楚，以说服读者，特别是学术期刊的审稿人。在这个过程中，要把握表述的客观性，特别是对问题的重要性和已有研究成果的评论，不要过于主观。在自己拿不准时尽可能地进行客观表述，避免不正确或不确切的主观评价。特别要避免使用过分夸大或贬低别人已有研究结论及问题的重要性的表述，例如"某某人第一次提出""这是某某领域最重要的公开问题""某某（人名）猜想"。我的意思是这些词不是不可以用，对那些业界有相当大共识的问题和结果，当然可以用。但对研究生同学初次撰写论文来说，论文研究的问题很可能（这里是假设，当然不排除你可能做了一件伟大的工作）是某一研究领域的重要性较为一般的问题。如果此时你在论文中使用上述词语，你的原意是想强调你的研究的重要性，但往往适得其反，给人一种拉大旗作虎皮的不好印象。总之，如果能恰如其分地把问题研究的脉络讲清楚，非常有助于说服审稿人接受你的论文发表。另一方面，从事学术研究的也是人，而人与人之间对同一问题有不同看法是很普通的事。在学术期刊处理稿件过程中，偶尔会遇到两个审稿意见大相径庭的情况。所以有些特别严格的审稿制度要求三份或更多份审稿意见，以决定稿件是否录用。

论文作者要做的第三件事，就是清晰、严谨地表述自己的研究结果。一个研究结果，通常是作者花费很多心血得到的。以论文的形式在专业期刊上发表出来，是一种被同行的认可。但你如果想得到这种认可，却也并非易事。首先，你的论文选题要被认可。有的同学可能苦思冥想，找到自己的研究题目，并得到了一些成果，但投出的论文却被杂志拒稿，很挫士气。其中可能的一个原因，是你的选题有问题。怎样选题，我想在本文后面单独讨论。除去研究选题的问题之外，另一个被杂志拒稿的可能原因就是你对自己的研究结果表述不够清晰严谨。其实，同学们在本科阶段必修的一些基础课程中，例如《数学分析》和《高等代数》，其中一个重要教学目标就是要学生学会如何清晰、严谨地表达数学论证。但很多同学不以为然，对如何表达不太重视，练习不够。到了写论文的时候，就会出现逻辑混乱不清晰，表达用词不够严谨的情况。在论文中表述自己的研究结果，有两种基本方式可供参考：一是主要结果叙述在先，论述证明在后；另一种是表述逐层深入，按照一定的逻辑关系发展，表述结果在后。其中容易出现的技术上的问题有：对一些数学符号的使用混乱，包括对论文中使用的一些符号不做说明，认为读者自己应该知道符号的含义，其实不尽然；对引用的已有的工具性结论（如某些预备性结论、不等式等）不标明出处。过多的符号不说明，使用混乱，会大大降低论文的可读性，会导致读者（特别是审稿人）的不适，降低稿件被接受的可能性。而对使用的已有结果出处没有很清楚的说明，轻则使得审稿人不得不花精力去核实你所用的结论，当审稿人失去耐心后，导致你的稿件被拒或延迟接受时间，重则被怀疑违反学术规范（但我相信你并非故意）。

如果能够做好上述三件事，一篇论文就具备了基本要素。建议同学们在撰写论文时可先构思论文结构，按照上述要素写下论文撰写提纲，然后再具体完成细节。俗话说"依葫芦画瓢""照猫画虎"，在第一次撰写论文时，可求助于老师和同学，让他们推荐几篇大家认可的论文，揣摩一下这些论文的构架，作为你的论文构架的参考。当完成论文初稿，并考虑确定向哪本学术期刊投稿后，还要查找该学术期刊对稿件的要求（通常可在期刊网站上找到），有些期刊有自己的稿件模板和版面要求。

撰写学位论文（硕士论文或博士论文）与投稿期刊的论文还有些不同。期刊论文往往是对单一问题的研究结果，学位论文（特别是博士论文）很多时候是对几个或一系列问题的系统研究结果，所以在论文的整体结构上更要安排合理。投给期刊的论文由于版面的限制通常详略要更为得当，不宜篇幅过长。但学位论文可以更详尽一些，如必要，也可以增加一些章节和附录，以达到论文的完整性和可读性。但学位论文在前述的论文要做到的三件事上与投稿期刊的论文基本是一样的。

到目前为止与同学们讨论的，主要是撰写论文的一些原则和技术性问题。一篇专业论文除了注意上述诸方面外，更重要的是论文研究成果的重要性。最终决定论文价值的应该是研究结果。但对于初次撰写论文的研究生同学，如果上述原则把握得不好，可能把一个尚有一些价值可发表的研究结果写成了糟糕的稿子，投出去后被拒或多次修改历尽坎坷延误发表。说实话，即使在科研上有所积累的人，都不能保证每篇稿子都那么令人满意，往往论文发表后自己再读也总觉得不如意。更何况论文要发表，总要经受期刊审稿，而处理稿件的期刊编委及其审稿人对稿件的看法有时不可避免地受偶然因素的影响，进而影响稿件命运。作为投稿人，我们只能尽自己最大的努力，让稿子在细节上尽可能完善，争取最好结果。

最后，我们谈谈论文研究问题的选取。作为研究生，论文选题通常依赖于指导教师，但也有少部分同学自己独立选题撰写论文。我主张鼓励研究生在可能的情况下独立选择论文研究题目。我的理由是，在研究生阶段，主要目标是学习做科研，而选择论文研究题目，是科研工作的重要组成部分。在研究生之前的学习阶段，主要经历集中于读书学习，从而，很多人习惯于老师出题，自己答题这样的模式。到了撰写论文阶段，仍然等着老师给题目，自己的任务就是执行老师交给的任务，完成老师提出的研究课题。这样也可能很平顺地完成研究生阶段的工作。毕业后，离开导师独立开展科研工作时，就会陷入找不到可以用来产生论文的科研题目、在科研上长时间止步不前的问题，最后失去科研的续航力，逐步脱离科研。

那么，研究生如何才能获得独立选择科研问题的能力呢？更具体点说，怎样选择一个可以研究并撰写科研论文的题目呢？首先要发掘自己感兴趣的问题。为做到这一点，要做到勤奋，多读、多听、多交流。多读，就是在导师的指导下，去阅读一定数量的文献资料。如何阅读文献，其实也是需要学习的。我认为阅读文献可分为两种情况：为了了解某一课题的研究状况而进行的泛读。在这种情况下，通常通过读文献的标题、摘要或简略浏览文献本身，了解文献研究的问题和主要结果，判断一下文献对自己感兴趣的科研领域的重要性。如果感觉重要，就会把这篇文献挑选出来，将来进一步细读。另一种，就是对已经挑选出的感兴趣的论文，逐篇细读。对于研究生同学，后一种情况可能更为常见。通常是导师或其他人推荐给你几篇围绕某一课题的文献。对于其中重要文献，你需在阅读过程中仔细分析，不停地向自己提问。例如：这篇文献对已有问题的处理结果依赖哪些条件？这些条件是否都合理？有没有可以弱化这些条件的可能？能否用新的方法替代文献中的方法？等等。经过这样的思考和对问题的逐一回答，往往会产生你的研究目标。也可能这些问题的解决构不成一篇高质量的论文，但是你的研究可能就此开启，逐渐积累，得到足以写成论文发表的研究结果。

上面的讨论只是我个人的一点体会。真正写出好的论文，还需同学们的卓越才华，同时也需要埋头苦干、不畏困难和勇攀学术高峰的精神。文中如有不当之处，愿意接受读者指教。

在 *IMA J. Numer. Anal.* 发表文章的经历

何银年（西安交通大学）

各位同学：大家好！我是西安交通大学数学与统计学院一名教师，一直从事定常和非定常 Navier-Stokes 方程组的有限元方法的研究。2007 年在中国科学院应用数学研究所访问时开始接触到非定常 MHD（磁流体动力学）方程组的有关知识，MHD 方程组就是 Navier-Stokes 方程组和 Maxwell 方程组进行耦合的时间依赖的非线性方程组，其难度可想而知。我的目标是进行 MHD 方程组全离散有限元半隐格式的无条件约束的收敛性数值分析工作，在以往的文献中关于非定常 Navier-Stokes 方程组和 MHD 方程组的全离散有限元半隐格式，都要求时间步长依赖于空间网格尺度的 CFL 约束条件，因此自己的目标具有一些挑战性的难度。经过三年的艰苦工作和探索，2010 年文章初步形成，由于自己初次进入这个研究领域，没有广泛的认可度，投稿文章多次被拒。然而这时自己有信心相信自己工作的价值，被拒稿后继续修稿，最后在 2013 年收到 *IMA J. Numer. Anal.* 杂志的修改意见，然后自己认真修改，逐点逐句回答每位审稿人的评审意见，最后在 2014 年文章终于被接受，在 2015 年见刊发表（从开始写作到发表长达 8 年之久），2016 年至今为 ESI 高被引论文。

我发表文章的体会是：（1）选准好的研究内容和目标，对自己的目标要有信心坚持不懈地去完成；（2）不怕被拒稿，反复修改后再选择合适的杂志进行投稿，即使文章被拒，有些审稿人的合理建议也要吸收学习，然后修改完善；（3）珍惜每次修改的机会，精益求精认真修改，然后逐点逐句回答评审人的意见。只要做到以上三点，你的文章就可能被发表在合适的期刊上。

下面我附上投稿经历和修稿过程，供大家阅读和参考。该文最终发表的信息如下。

He，Yinnian，Unconditional convergence of the Euler semi-implicit scheme for the three-dimensional incompressible MHD equations. IMA J. Numer. Anal. 35 (2015)，no.2，767 - 801. 最初投稿编号：ID IMAJNA -AI - 2013 - 001。

两位评审人的意见如下。

- **Referee 1（IMAJNA - AI - 2013 - 001）**

Comment 1：Page 5，line 41，Assumption（A2）. Add some comments on why you use this assumption. I mean, the assumption A1 was justified by saying that you rather assume the existence of the solution with regularity than assume small initial data. So how do you justify assumtoin A2? Is it reasonable to assume A2? Do we still have soma family of data f and g such that A2 is satisfied? Aren't we speaking about an empty set?

Comment 2：What are the requirements for regularity of functions μ, σ? Can they be discontinuous? Or even vanishing? Please add a remark on this. E. g., in L. Banas, A. Prohl, Converget finite element discretization of the multi-fluid nonstationary incompressible magnetohydrodynamics equations, MATHEMATICS OF COMPUTATION, Volume 79, Number 272, October 2010, Pages 1957 - 1999, the authors studied multi-fluids, where density, viscosity and conductivity are allowed to be discontinuous. Also, in S. Durand, I. Cimrak, P. Sergeant, Adjoint variable method for the time-harmonic Maxwell equations. COMPEL Vol. 28(5)（2009），pp.1202 - 1215, the authors studied quasistatic Maxwell equations where conductivity is allowed to vanish. Please, mention these works as references.

Comment 3：The definition of weak solutions is strange. I mean, what does it mean u_t? In relations (2.1)-(2.3) there should be an integral over the time, right? You define variational formulation of the problem by: For all $t \in [0, T)$ find $(u, p, H) \in X \times M \times W$ such that for all test functions $(v, q, B) \in X \times M \times W$ the relations (2.1)-(2.3) are satisfied. I suggest that you need to rewrite the definition of weak solutions, because in state as it is now, it does not give any sense. Even the work [31] of 1983 has the weak definition through integrals over time — see relation (2.19) in [31].

Comment 4： Page 8, proof of Theorem 3.3. You say that you differentiate in time both sides of (2.1) and (2.2). How can you differentiate an equation, when you do not have any information about higher regularity of u, e. g., existence of u_{tt}?

Comment 5 - typos： Page 33, lines 16 - 17, Section 9. Conclusions, you have missing number of the paragraf (??) Page 33, lines 22 - 23, should by word "convergent" and not "convergence". Page 2 lines：32, 33, 36, 44—there are blank spaces missing mostly before " (". Page 2 line 46, should be "will be discussed" and not "will discussed".

Comment 6： Page 30, please, split the exhausting relations in several parts.

- **Referee 2 (IMAJNA - AI - 2013 - 001)**

The main claim in this article is "To the best of our knowledge, this seems to be the first time to establish (sic) the $H^1 - L^2$ unconditional convergence of the discrete solution (u_h^n, p_h^n, H_h^n) to a linearized finite element system of the nonstationary MHD equations in three dimensions." Perhaps some extra care could be taken over the language for the claim. This is not the only piece of awkward prose in the article as there are several places where the English needs to be improved.

This is a generally well written paper that appears to indeed establish this claim, although I have not gone through every step of every deviation. The only major criticisms I have are related to the way that the highly technical results are presented. For instance the thicket of inequalities given on pages 18, 24, 26, 29, 30, and 32 (and indeed elsewhere) are devoid of useful textual explanation. Since various steps in the derivations go without comment it seems perhaps these technical results could be relegated to an appendix for the detail oriented reader who is willing to go through the process of understanding how to rederive these results. Alternatively the author could help the general reader understand the analytical process and break up the thicket of inequalities with some comments that help walk the reader through the densely presented proofs. In the current state large swaths of the analysis are almost unreadable to all but a few experts.

The technical Lemma 2.1 is rather trivial and seems to not require a proof. Finally, I was disappointed to see a lack of computational results. However, overall the paper is interesting, and the unconditional stability result alone makes the article worth publishing. Typos: p12 and p33 there is a broken reference and?? appears instead.

认真研究评阅人的意见后，我尽心尽力地修改。下面是写给主编的信，其中逐条回答每一位审稿人的意见。

Dear Prof. Arieh Iserles,

We are very pleased to learn from your letter about revision for our manuscript (ID IMAJNA-AI-2013-001). Thank you for your comments and advice which not only are valuable and very helpful for revising and improving our manuscript but also have important guiding significance to our research. We have studied the comments carefully and have made corrections which we hope will meet with approval.

- **Responses to Referee 1**

We are truly grateful to your critical comments and thoughtful suggestions on our manuscript. Based on these comments and suggestions, we have made careful modifications on the original manuscript.

Comment 1: Page 5, line 41, Assumption（A2）. Add some comments on why you use this assumption. I mean, the assumption A1 was justified by saying that you rather assume the existence of the solution with regularity than assume small initial data. So how do you justify assumption A2? Is it reasonable to assume A2? Do we still have soma family of data f and g such that A2 is satisfied? Aren't we speaking about an empty set?

Answer 1: *We add a remark after Assumption（A2）: The validity of Assumption（A2）is known（see [2, 7, 9, 23, 33]）if f, $g \in L^2$ $(\Omega)^3$ and the boundary of Ω is of C^2, or it is a convex polyhedra. We shall make frequent use of Assumption（A2）for the regularity estimates of $(u(t), p(t), H(t))$.*

Comment 2: What are the requirements for regularity of functions μ, σ? Can they be discontinuous? Or even vanishing? Please add a remark on this. E. g., in L. Banas, A. Prohl, Converget finite element discretization of the multi-fluid nonstationary incompressible magnetohydrodynamics equations, MATHEMATICS OF COMPUTATION, Volume 79, Number 272, October 2010, Pages 1957–1999, the authors studied multi-fluids, where density, viscosity and conductivity is allowed to be discontinuous. Also, in S. Durand, I. Cimrak, P. Sergeant, Adjoint variable method for the time-harmonic Maxwell equations. COMPEL Vol. 28 (5) (2009), pp.1202–1215, the authors studied quasistatic Maxwell equations where conductivity is allowed to vanish. Please, mention these works as references.

Answer 2: *According to the Referee's valuable advice, we add a remark on page* 1: *Here we assume that* ν, μ *and* σ *are positive constants. For the case of* ν, μ, σ *and the density being allowed to be discontinuous positive functions on* $x \in \Omega$, *the reader can refer to the paper by Banas and Prohl* [2]. *Also, Durand, Cimrak and Sergeant studied the quasistatic Maxwell equations where* σ *is allowed to vanish in* [8].

Comment 3: The definition of weak solutions is strange. I mean, what does it mean u_t? In relations (2.1)–(2.3) there should be an integral over the time, right? You define variational formulation of the problem by: For all $t \in [0, T)$ find $(u, p, H) \in X \times M \times W$ such that for all test functions $(v, q, B) \in X \times M \times W$ the relations (2.1)–(2.3) are satisfied. I suggest that you need to rewrite the definition of weak solutions, because in state as it is now, it does not give any sense. Even the work [31] of 1983 has the weak definition through integrals over time—see relation (2.19)in [31].

Answer 3: *Here the weak solution* $u \in L^2(0, T; X_0)$ *or* $u(t) \in X_0$ *for almost all* $t \in [0, T)$ *and* $u_t \in L^2(0, T; X'_0)$ *or* $u_t(t) \in X_0$ *for almost all* $t \in [0, T)$. *Yes, in relations* (2.1) – (2.3) *there should be an integral over the time. But, relations* (2.1) – (2.3) *is also true by adding* **for almost all** $t \in (0, T)$, *see relation* (2.19) *in* [33] (*old* [31]) *in the case of the MHD equations or see relation* (2.9) *in* [23] (*old* [21]) *in the case of the Navier-Stokes equations.*

Comment 4: Page 8, proof of Theorem 3.3. You say that you differentiate in time both sides of (2.1) and (2.2). How can you differentiate an equation, when you do not have any information about higher regularity of u, e. g., existence of u_{tt}?

Answer 4: *From Theorem* 3.2, *we have* $u, H \in L^2(0, T; (H^2)^3)$. *Thus,* $(u(t), H(t))$ *is the strong solution. Thus, we can differentiate* *in time both sides of* (2.1) *and* (2.2). *The same manner was used in relation* (3.9) *of* [26] (*old* [24]) *by Hill and Sull and in relation* (2.10) *of* [23] (*old* [21]) *by Heywood and Rannacher. From new relations* (3.11)–(3.12), *we can prove that* u_t, $H_t \in L^2(0, T ; (H^1)^3)$ *and* u_{tt}, $H_{tt} \in L^2(0, T ; (H^{-1})^3)$.

Comment 5-typos：1. Page 33, lines 16–17, Section 9. Conclusions, you have missing number of the paragraf (??).

2. Page 33, lines 22–23, should by word "convergent" and not "convergence".

3. Page 2 lines：32, 33, 36, 44—there are blank spaces missing mostly before "（".

4. Page 2 line 46, should be "will be discussed" and not "will discussed".

Answer 5：*We have revised the above typos mistakes.*

Comment 6：Page 30, please, split the exhausting relations in several parts.

Answer 6：*We have revised some inequalities by simple manner.*

· **Responses to Referee 2**

We are truly grateful to your critical comments and thoughtful suggestions on our manuscript. Based on these comments and suggestions，we have made careful modifications on the original manuscript.

Comment 1：The main claim in this article is "To the best of our knowledge, this seems to be the first time to establish (sic) the $H^1 - L^2$ unconditional convergence of the discrete solution (u_h^n, p_h^n, H_h^n) to a linearized finite element system of the nonstationary MHD equations in three dimensions." Perhaps some extra care could be taken over the language for the claim. This is not the only piece of awkward prose in the article as there are several places where the English needs to be improved.

Answer 1: *We have revised this main claim and some other poor English statements in this article.*

Comment 2: This is a generally well written paper that appears to indeed establish this claim, although I have not gone through every step of every deviation. The only major criticisms I have are related to the way that the highly technical results are presented. For instance the thicket of inequalities given on pages 18, 24, 26, 29, 30, and 32 (and indeed elsewhere) are devoid of useful textual explanation. Since various steps in the derivations go without comment it seems perhaps these technical results could be relegated to an appendix for the detail oriented reader who is willing to go through the process of understanding how to read these results. Alternatively the author could help the general reader understand the analytical process and break up the thicket of inequalities with some comments that help walk the reader through the densely presented proofs. In the current state large swaths of the analysis are almost unreadable to all but a few experts. The technical Lemma 2.1 is rather trivial and seems to not require a proof.

Answer 2: *Yes, some inequalities are too thicker so that some readers are unwilling to read. We have revised some inequalities by more simple manner so that the 34 pages of old version are changed into the 31 pages of the new version. Also, the proof of Lemma 2.1 is deleted.*

Comment 3: Finally, I was disappointed to see a lack of computational results.

Answer 3: *I am sorry that no computational result was given in the paper. In this paper, our main purpose is to establish the unconditional stability and convergence of the semi-implicit scheme, and including numerical results will not improve the paper and will result in a paper which is overly long.*

Especially, in the Acknowledgements, we added our thanks to editor and anonymous referee for their helpful suggestions on the quality improvement of our present paper.

Once again, we appreciate the editor and anonymous referee's warm work earnestly, and hope that the corrections will meet with approval. We look forward to your information about our revised paper and thank you for your good comments.

Sincerely yours,

Yinnian He

文章本身是精心准备的，修改也是尽最大努力去做的。功夫不负有心人，论文最终被接受发表，现已成为 ESI 高被引论文。

最后祝大家学业有成，科研成果丰硕！

一篇计算数学论文及其
相应的研究过程

伍渝江（兰州大学）

一、引言

我国文化传统历来重视教育、重视人才培养。时至今日，教育已成事关国家发展、事关民族未来之事业，是国之大计、党之大计。我们肩负着落实立德树人的重任，新时代需要我们进一步解放思想、改革创新、奋发图强，加快推进中国特色世界一流大学和一流学科建设。在此背景下，母校上海大学编撰《数学科学与技术研究论文写作：案例与方法》，无疑是弘扬科学精神、加强数学教育、传播数学文化、推进数学创新、培养一流人才的重要举措。我作为一名普通的数学教育工作者，能于此谈谈自己的点滴经历，抛砖引玉，倍感荣幸。

数学研究是人类科学研究中充满智慧的活动，每段研究历程皆可视作智慧之旅。

早期的历史表明，数学起源于人类的生产和生活活动。经过数千年的社会生产实践和科学活动的推动，人类历史已将数学发展成为一门庞大的科学体系，它成为了人们研究现实世界的空间形式和数量关系的科学，它将研究数量、结构、变化、空间以及信息等诸多概念。数学理论的形式往往具有高度的抽象性，而实质上它总是扎根于现实世界，因此可以广泛地应用于自然科学和技术的各个部门。数学知识对改造世界的实践起着重要的、关键的作用，利用数学手段，人们可以严格描述一些事物的抽象结构和组成模式，找到事物的共性，从而普遍应用于现实世界中各种表面形态迥异的事物。实际上人类历史发展和社会生活中，数学发挥着不可替代的作用，它成为学习和研究现代科学技术必不可少的基本工具。

作为数学学科的古老分支之一的计算数学自二次世界大战之后以超常的规模发展，得益于各种高性能计算机的不断问世，从根本上提升了功能强大的计算工具，科学计算的格局和面貌焕然一新。现代计算数学的重要任务之一就是要研究在计算机上进行大规模计算的有效算法及其相应的数学理论，并伴之以恰当的数值实验和数据分析。计算数学和计算机一起已成为众多领域研究工作开展不可或缺的工具与手段，计算早已成为人们从事科学研究的第三种重要手段。

自然科学和工程技术中许多基本规律往往以微分方程的形式表示，建立起各种数学模型。而能求出精确解的方程只有极少部分，因而只有借助于计算数学的办法，这就是微分方程数值解面临的主要问题。依托于各种成熟的或新创的离散格式与技巧，离散后的微分方程通常将得到的是线性或非线性方程组，从而有效求解线性或非线性方程组的数值代数研究就成为了基本和重要的问题，它实际上也成了众多学者呕心沥血钻研的长盛不衰的科学问题。囿于我们的学识，我将以此领域的一篇论文为主要案例，简介我们的研究方法、研究结果和研究过程。这篇论文的信息如下：

Ai-Li Yang, Yang Cao, Yu-Jiang Wu, Minimum residual Hermitian and skew-Hermitian splitting iteration method for non-Hermitian positive definite linear systems. *BIT Numerical Mathematics*，59 (2019)，299 – 319. [MR3921381]

二、研究背景

许多科学计算问题需要求解线性代数方程组 $Ax = b$，其中系数矩阵 A 是正定的。当 A 是结构化的大型稀疏阵，求解方程组的迭代法比直接法更引人入胜，因为直接法的代价高。进一步，由于舍入误差的积累，直接法可能不稳定。最近几十年，克雷洛夫子空间方法作为一种迭代法获得广泛关注。尽管克雷洛夫子空间方法有很好的收敛性质，但所需的存储空间总随着问题的规模增长很快。所以，这篇论文的研究关注另一类迭代方法，即矩阵分裂迭代法。

矩阵 A 可以有经典的唯一的埃尔米特和反埃尔米特分裂（或称 HS 分裂）$A = H(A) + S(A)$，其中 $H(A)$ 和 $S(A)$ 分别表示埃尔米特部分和反埃尔米特部分。2003 年，世界著名计算数学专家、美国三院院士 G. 戈卢布等人据此建立了目前堪称经典的 HSS 迭代法。这个方法有几种数学上等价的不同的表示形式，其中一种是用残量的形式来表示的。即，

$$\begin{cases} x^{(k+1/2)} = x^{(k)} + (\alpha I + H(A))^{-1}(b - Ax^{(k)}), \\ x^{(k+1)} = x^{(k+1/2)} + (\alpha I + S(A))^{-1}(b - Ax^{(k+1/2)})。 \end{cases}$$

经人们研究发现，这些等价形式若论实际计算效能，又以残量表示的形式为佳。

由于 HSS 迭代法的高效性和稳健性，近年来 HSS 的许多变形被提出以求解各种不同形式的线性方程组。例如，预处理 HSS 迭代法和加速 HSS 迭代法用于求解鞍点问题；修正 HSS（即 MHSS）迭代法及其预处理变形用于求解复线性方程组或其等价的实形式；HSS 迭代法及其参数化版本用于求解奇异问题；当用作求解器时，HSS 方法及其变形也可用作预处理子去加速克雷洛夫子空间方法的收敛性。

虽然近年发表了大量的 HSS 型迭代方法，大多数都是集中于应用这些方法求解不同类型的问题，特别是应用于鞍点类型问题。只有很少论文通过研究 HSS 迭代格式本身来提出一些加速的 HSS 型方法，例如使用超松弛技巧加速 HSS 方法的收敛性的 SOR 加速 HSS 迭代法（SOR-HSS）。此外，因为埃尔米特部分和反埃尔米特部分具有相当不同的性质，人们也在 HSS 迭代格式中分别对 $H(A)$ 和 $S(A)$ 使用不同的位移参数以加速 HSS 的收敛性，这导致产生广义预处理 HSS 迭代法。使用试验获得最优参数，两种加速方法都比原来的 HSS 迭代方法有效。不过在实施两种方法的每一种时，我们都要确定两个迭代参数的值，这是一件费力又耗时的事情。

借鉴于鞍点问题求解研究中的非定常 Uzawa 方法的加速技巧，不同于所有现有的 HSS 的变形，我们将对残量形式的 HSS 迭代格式通过引入两个更多的控制参数而介绍一种非定常 HSS 迭代格式。因为新格式的新控制参

数是通过极小化对应的残量范数来确定的，我们就称这种非定常 HSS 迭代格式为极小残量 HSS 迭代法（简称 MRHSS 迭代法）。

三、MRHSS 迭代法及其性质

通过修正 HSS 迭代格式的残量形式来推导 MRHSS 迭代法。这种形式中可形成两个半步的搜索方向，分别表示从第 k 步到第 $k+1/2$ 半步的搜索方向和第 $k+1/2$ 半步到第 $k+1$ 步的搜索方向。于是 HSS 迭代格式的残量形式二式都沿搜索方向前进单位步长。为进一步改进格式的有效性，提出以下两个有趣的问题：

——前进单位步长是否最佳的？

——若非最佳的，能否找到最佳步长？

为回答这两个问题，引进两个任意参数 t 和 c 以控制步长，因是复方程组，t 和 c 取自复数域。若对两个半步的残量在 2 范数下取极小化，获得关于 $\mathrm{Re}(t)$，$\mathrm{Im}(t)$，$\mathrm{Re}(c)$，$\mathrm{Im}(c)$ 四个变量的极小值点。不过，由于涉及一些未知矩阵的 HS 分裂，这四个变量的极小值点实际上是求不出来的。所幸，直接推导表明以下复数表达式更容易获得：$t = \mathrm{Re}(t) + i\mathrm{Im}(t)$，$c = \mathrm{Re}(c) + i\mathrm{Im}(c)$。

自此，MRHSS 迭代法的算法可以顺利设计了。主要步骤为：

给定初始猜测 $x^{(0)} \in \mathbb{C}^n$，计算 $\{x^{(k)}\}$ 直至收敛（$k = 0, 1, 2, \dots$）

(i) $x^{(k+1/2)} = x^{(k)} + t\delta^{(k)}$，其中

$$t = (r^{(k)}, A\delta^{(k)}) / \|A\delta^{(k)}\|^2, \quad \delta^{(k)} = (\alpha I + H(A))^{-1} r^{(k)}。$$

(ii) $x^{(k+1)} = x^{(k+1/2)} + c\delta^{(k+1/2)}$，其中

$$c = (r^{(k+1/2)}, A\delta^{(k+1/2)}) / \|A\delta^{(k+1/2)}\|^2, \quad \delta^{(k+1/2)} = (\alpha I + S(A))^{-1} r^{(k+1/2)}。$$

注 1 参数 t 和 c 均依赖于迭代指标 k。

注 2 如果选取 $t = c = 1$，则 MRHSS 迭代法退化为古典的 HSS 迭

代法。

注 3　通过简单分析，使用 HSS 预处理子，可得到所谓预处理 Orthomin(2)方法，简记为 POM(2)，易知 POM(2)与 MRHSS 是两个完全不同的方法。

注 4　MRHSS 迭代法不是标准的 Krylov 子空间方法。

由算法设计可知：

$\mathrm{Re}(t)$，$\mathrm{Im}(t)$是第 k 步到第 $k+1/2$ 半步的残量在 2 范数下的极小点。

$\mathrm{Re}(c)$，$\mathrm{Im}(c)$是第 $k+1/2$ 半步到第 $k+1$ 步的残量在 2 范数下的极小点。

一个重要的问题：四元组 $\mathrm{Re}(t)$，$\mathrm{Im}(t)$，$\mathrm{Re}(c)$，$\mathrm{Im}(c)$是第 k 步到第 $k+1$ 步的残量在 2 范数下的极小点吗？

为了回答此问题，我们建立了一个埃尔米特矩阵二次型关于向量范数平方的形式偏导数和该埃尔米特矩阵二次型本身的关系式，并借助于这个关系式获得了上述问题的肯定答案。进一步明白了由此定义的 t 和 c 实际上在复数域达到了最优。

至于 MRHSS 方法的收敛性条件的获得，主要依赖于一些复矩阵的非零复向量 Rayleigh 商值域的分析。我们找到了与 HS 分裂有关的两类复矩阵，即

$$M = A(\alpha I + H(A))^{-1} \text{ 和 } N = A(\alpha I + S(A))^{-1},$$

则矩阵 M 的非零复向量 Rayleigh 商值域和矩阵 N 的非零复向量 Rayleigh 商值域都不含 0 元素的情形下，对任意初始向量都能保证 MRHSS 迭代法收敛，并且能确定残量范数的压缩因子。

四、对流扩散的数值算例

我们曾用两个例子来检验 MRHSS 迭代法的有效性。分别在迭代步数（IT）和计算机耗时（CPU）两方面比较数值结果，参与比较的现有方法为：HSS，POM（2），PGMRES（预处理极小残量法）和 SOR-HSS。用于

POM(2)和 PGMRES 的预处理子都是 HSS 预处理子。

HSS, POM(2), PGMRES 和 MRHSS 四种方法中的单参数与 SOR-HSS 中的双参数均选为实验中找到的最优参数，这样可以导致最少迭代步数。如果最优迭代参数形成一个区间，则我们选的最优参数既属于该区间且耗用最小的 CPU 时间。

如所知，使用实验最优参数之目的是检验所用迭代法的最优执行性能，然而在实际应用中使用实验最优参数往往是不现实的。于是，许多研究者试图讨论 HSS 迭代法或 HSS 预处理子的最优参数。以下我们列出三种参数选择，并保持在原文中的标号（33），（34），（35）：

$$\alpha^* = \sqrt{\lambda_{\max}(H(A))\lambda_{\min}(H(A))}, \tag{33}$$

$$\alpha_H = \arg \cdot \min_{\alpha>0}\{\|(\alpha I - H(A))(\alpha I - S(A))\|_F\}, \tag{34}$$

$$\alpha_e = (\|H(A)\|_F + \|S(A)\|_F)/(2n)。 \tag{35}$$

我们将分别使用（33），（34），（35）定义的参数值来检验四种迭代方法（即 HSS, POM(2), PGMRES, MRHSS）的计算效能。

实际计算中，迭代方法的初始猜测均取为 $(1, 1, ..., 1)^T$。迭代过程终止的条件是当前残量满足 $<10^{-6}$ 或者迭代步数超过 5 000。此外，所有计算均按双精度用 MATLAB（版本为 9.1.0.441655（R2016(6)））实现，计算环境为 PC 机 3.10 GHz 中央处理单元［Intel（R）Core（TM）i7－4770S］，8.00 GB 内存。

例 1 考虑如下二维对流扩散方程：

$$-\Delta u + a(x, y)u_x + b(x, y)u_y = f(x, y), \text{ in } \Omega,$$
$$u = 0, \qquad \text{on } \partial\Omega,$$

其中 Ω 是单位正方形。系数函数 $a(x, y)$, $b(x, y)$ 选为

Case I. $a(x, y) = x\sin(x+y)$, $b(x, y) = y\cos(x, y)$,

Case II. $a(x, y) = 5ye^{xy}$, $\qquad b(x, y) = 5xe^{x+y}$。

使用步长 $h = 1/l$ 的五点中心差分格式去离散上述问题，获得了一类线

性方程组 $Ax=b$。

于是，所有五个检验迭代方法（HSS，SOR-HSS，POM(2)，PGMRES，MRHSS）都用于求解这类线性方程组。在表 1 中，我们列出了不同系数函数与不同网格大小情形下迭代方法 HSS，POM(2)，PGMRES，MRHSS 的单个实验最优参数和 SOR-HSS 迭代方法的两个参数。

使用两个实验最优参数，表 2 中我们列出了五种迭代法在两种情形的系数函数和三种不同网格大小（即 $l=40, 80, 160$）时的数值结果，仍显示迭代步数和耗时 CPU。

由表 2 中所列数值结果我们可以看到，对不同的系数函数和不同的网格大小，具有各自最优参数的五种检验的迭代函数始终收敛。SOR-HSS 和 POM(2)这两种迭代方法比 HSS 迭代方法表现得更好。然而，它们的计算效能低于 PGMRES 方法。五种检验的迭代法中，MRHSS 始终是最有效的一个，它花费较少的迭代步数和计算时间就收敛到解了。进一步，MRHSS 迭代法花费的迭代步数几乎是 h-独立的。

除了使用实验最优参数，我们还采用了三类由(33),(34),(35)确定的容易计算的参数值来进一步检验五种迭代方法的有效性。参数值列于表 3，使用这些参数值的检验迭代方法的数值结果列于表 4。由数值结果我们可以看到，相较于 HSS 和 POM(2)两种迭代法，另两类迭代法 PGMRES 和 MRHSS，无论采用三类参数值选取的哪一类，始终在最大迭代步数内达到收敛。虽然在某些情形，PGMRES 方法比 MRHSS 迭代法略胜一筹，但总体状况考虑，MRHSS 迭代方法却是最稳健的方法。

Table 1　The experimentally found optimal parameter
values of the tested methods for Example 1

	$l=40$	$l=80$	$l=160$
Case I			
HSS	0.338	0.187	0.103
SOR-HSS	0.052 \ 0.841	0.026 \ 0.901	0.010 \ 0.900
POM(2)	0.031	0.015	0.006
PGMRES	0.065	0.023	0.009
MRHSS	0.000 8	0.000 2	0.000 1

	$l = 40$	$l = 80$	$l = 160$
Case II			
HSS	0.382	0.208	0.113
SOR-HSS	0.286 \ 0.954	0.144 0 \ 0.952	0.073 \ 0.951
POM(2)	0.288	0.132	0.057
PGMRES	0.652	0.420	0.192
MRHSS	0.047	0.009	0.003

Table 2　The numerical results of the tested methods with experimentally found optimal parameter values for Example 1

Method	$l = 40$		$l = 80$		$l = 160$	
	IT	CPU	IT	CPU	IT	CPU
Case I						
HSS	116	0.078 0	219	2.355 6	407	34.149
SOR-HSS	26	0.031 2	38	0.436 8	50	4.149 6
POM(2)	22	0.031 2	26	0.405 6	36	2.901 6
PGMRES	15	0.031 2	18	0.265 2	24	2.028 0
MRHSS	3	0.015 6	3	0.015 6	3	0.312 0
Case II						
HSS	91	0.078 0	171	1.778 4	314	26.177
SOR-HSS	74	0.062 4	132	1.341 6	225	18.673
POM(2)	77	0.062 4	135	2.043 6	222	18.549
PGMRES	38	0.031 2	58	1.014 0	88	8.080 9
MRHSS	29	0.015 6	26	0.608 4	22	2.184 0

Table 3　The values of parameter α computed by formulas (33), (34) and (35) for Example 1

	(33)	(34)	(35)
Case I			
$l = 80$	0.155 1	8.577 5E−6	0.028 7
$l = 160$	0.077 1	2.140 9E−6	0.014 2
Case II			
$l = 80$	0.137 8	2.286 5E−3	0.029 3
$l = 160$	0.068 5	5.739 2E−4	0.014 3

Table 4 **The numerical results of iteration methods with different values of parameter α for Example 1**

Method		(33) IT	(33) CPU	(34) IT	(34) CPU	(35) IT	(35) CPU
Case I							
$l = 80$	HSS	260	2.870 4	—	—	1 131	11.872
	POM(2)	51	0.795 6	—	—	34	0.499 2
	PGMRES	29	0.468 0	746	70.669	19	0.343 2
	MRHSS	134	1.576 3	3	0.031 2	32	0.499 2
$l = 160$	HSS	541	49.515	—	—	2 373	206.61
	POM(2)	77	7.300 8	—	—	49	4.726 8
	PGMRES	39	3.962 4	1 528	740.27	25	2.215 2
	MRHSS	244	24.383	3	0.234 0	53	4.976 4
Case II							
$l = 80$	HSS	235	2.808 0	—	—	753	7.831 3
	POM(2)	135	2.168 4	—	—	—	—
	PGMRES	70	1.294 8	1 298	277.98	196	5.054 4
	MRHSS	101	1.840 8	39	0.452 4	32	0.530 4
$l = 160$	HSS	478	44.616	—	—	1 547	137.94
	POM(2)	225	20.311	—	—	—	—
	PGMRES	99	9.750 1	2 726	3 347.8	224	26.317
	MRHSS	177	16.521	33	2.792 4	40	3.900 0

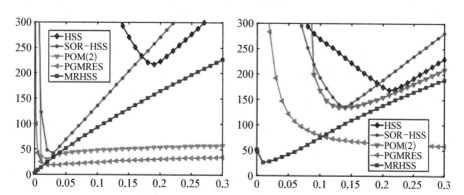

Fig.1 **Numbers of iteration steps versus iteration parameter α for the Cases I (left) and II (right) of Example 1**

在图 1 中，我们画出了 $l=80$ 时五种检验迭代方法关于各自迭代参数的迭代步数图。SOR-HSS 方法的另一个迭代参数的选取法则是：Case I 时取

0.901，Case II 时取 0.952（参见表 1）。数值曲线显示，在两种选择情形，MRHSS 迭代法始终比 HSS 和 SOR-HSS 迭代法耗费较少的迭代步数。而与 POM(2) 和 PGMRES 两种方法相比较，MRHSS 迭代法在 Case I 对参数值更敏感，即，它的计算效能大大地依赖于参数值。而对 Case II，MRHSS 迭代法比方法 POM(2) 远胜一筹。然而，它仍然比 PGMRES 方法对参数的值更敏感。此外，对两种情形，MRHSS 迭代法的一个显著的特征是，当参数趋于 0 时，它获得最少迭代步数。

五、抛物问题帕第逼近型积分的数值算例

例 2　来自于抛物问题帕第逼近型积分离散的复的 Hermitre 正定线性系统

$$\left(I + \frac{\tau}{4}(1 + 1/\sqrt{3})L\right)x = b,$$

其中矩阵 L 表示负的拉普拉斯算子标准五点差分离散。

在表 5 中，对五种检验迭代方法（HSS, SOR-HSS, POM(2), PGMRES, MRHSS）我们列出了具不同网格个数 l 时的实验最优参数值。使用这些最优值，五种迭代方法的数值结果列于表 6 中。数值结果表明，与 HSS, SOR-HSS, POM(2) 和 PGMRES 四种方法比较，MRHSS 迭代法始终花费最少迭代次数与计算时间就能收敛到解。进一步，就此例而言，MRHSS 迭代法的迭代步数是 l-独立或 h-独立的。

对此例，除了使用实验最优参数值，在我们检验五种迭代方法的数值有效性时，也使用了由（33），（34），（35）定义的参数值。由（33），（34），（35）定义的参数值在 $l=80$ 时为 12.987, 75.574, 0.897 7，而在 $l=160$ 时为 18.132, 151.15, 0.889 6。使用这些参数值，检验迭代方法的数值结果列于表 7。我们可以看到，无论使用什么样的参数值，PGMRES 和 MRHSS 这两种迭代法始终优于 HSS 和 POM(2) 这两种方法。然而，使用不同的参数值，则 PGMRES 和 MRHSS 各有所长。

Table 5   The experimentally found optimal parameter values of the HSS-type methods for Example 2

	$l = 40$	$l = 80$	$l = 160$
HSS	8.2	10.6	15.1
SOR-HSS	8.2 \ 1.1	12.0 \ 1.2	16.4 \ 1.2
POM(2)	5.1	8.3	9.8
PGMRES	5.6	8.7	11.1
MRHSS	0.21	0.24	0.31

Table 6   The numerical results of iteration methods with experimentally found optimal parameter values for Example 2

Method	$l = 40$		$l = 80$		$l = 160$	
	IT	CPU	IT	CPU	IT	CPU
HSS	36	0.046 8	50	0.795 6	69	7.254 0
SOR-HSS	31	0.031 2	38	0.561 6	50	5.179 2
POM(2)	19	0.015 6	24	0.483 6	30	3.338 4
PGMRES	20	0.031 2	26	0.530 4	33	4.056 0
MRHSS	5	0	5	0.124 8	5	0.577 2

Table 7   The numerical results of iteration methods with different values of parameter α for Example 2

Method	(33)		(34)		(35)	
	IT	CPU	IT	CPU	IT	CPU
$l = 80$						
HSS	58	0.686 4	336	4.383 6	571	7.659 6
POM(2)	29	0.452 4	101	1.794 0	126	2.121 6
PGMRES	27	0.405 6	41	0.998 4	50	1.123 2
MRHSS	27	0.402 8	98	1.762 8	6	0.156 0
$l = 160$						
HSS	81	8.002 9	670	68.172 4	1 105	110.14
POM(2)	37	3.962 4	190	19.125 7	232	23.385
PGMRES	35	3.868 8	58	6.786 0	75	9.126 1
MRHSS	37	3.946 8	196	20.654 5	6	0.655 2

在图 2 中，我们分别对 $l=80$ 和 $l=160$ 的情形，画出了五种检验迭代方法（HSS, SOR-HSS, POM(2), PGMRES, MRHSS）关于迭代参数的迭代

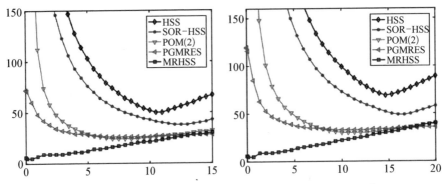

Fig.2 Numbers of iteration steps versus iteration parameter α for the cases $l = 80$ (left) and $l = 160$ (right) of Example 2

步数的图。而 SOR-HSS 的另一个参数在 $l=80$ 和 $l=160$ 两种情形都取为 1.2。与 HSS 迭代法和 SOR-HSS 迭代法相比较，POM(2) 迭代法和 PGMRES 迭代法始终花费较少的迭代步数而满足同样的停止准则。在五种检验的迭代法中，MRHSS 迭代法表现得最为有效，它几乎对所有的检验参数值都优于另外四种方法，且 MRHSS 迭代法的迭代步数对参数值不敏感。此外，类似于例1，当参数非常靠近0时，MRHSS 迭代法也达到其最少迭代步数。

六、结束语

由例 1 和例 2 的数值结果我们可以得出结论，这篇论文讨论的 MRHSS 迭代法，特别当参数很小的时候非常有效、非常稳健。

换言之，我们讨论了一类非 Hermite 正定线性方程组近似求解的 MRHSS 迭代法，一方面仔细分析了 MRHSS 迭代法的收敛性质，一方面用较为丰富的数值结果表明，MRHSS 迭代法是非常有效和稳健的。

如果要对本项研究做些梳理，以便对初学者有所启示，也许可以增添以下几句话来结束本文。

——本项研究的成文并发表曾得到数位专家的有益的改良建议，再次感谢！

——本项研究的创新思路在于直接针对 HSS 迭代法本身进行理论性改造，而非着眼于不同形式的应用模式。

——论文中证明了新设两参数真正使残量达到极小，获得了 MRHSS 迭代法独特的收敛性条件，这应该是我们研究中解决的关键科学问题。

——结合三种参数选择的数值实验结果与比较，有力地支撑了创新理论方法。

——论文研究的成功之处是建立了一类真正的非定常的 HSS 类迭代方法；不足之处或许在于只考虑了 MRHSS 迭代法的"精确"版本，未涉足其非精确变形的分析。

从"黎曼问题"谈起

李杰权(北京应用物理与计算数学研究所、
北京大学应用物理与技术中心)

伟大的德国数学家黎曼(Riemann)(1826—1866)生前只发表了九篇论文,创立了黎曼几何、复变函数论(创始人之一);他的黎曼猜想被认为是解析数论中最有影响的问题之一,推动着数学物理等领域的发展。本文从他在空气动力学中的伟大贡献——"黎曼问题"谈起,品味这一工作如何产生广泛而深刻的科学影响。

黎曼问题的数学表述如下:考虑非线性双曲守恒律组

$$u_t + f_x(u) = 0,$$

这里 $u(x, t)$ 是未知量,$f(u(x, t))$ 为非线性通量函数,t 是时间变量,x 为空间变量。这一方程组包括流体力学中质量、动量和能量守恒的三大定律。假定初始时刻 $t = 0$,$u(x, 0)$ 由两个不同的常状态 u_L,u_R 组成,

$$u(x, 0) = \begin{cases} u_L, & x < 0, \\ u_R, & x > 0, \end{cases}$$

则称这一初值问题为"黎曼问题"。它由空气动力学中的激波管实验提炼而出,黎曼首先给出数学分析,后人以他的名字来命名。

一、黎曼问题中的原始创新

黎曼问题出自于他的"论有限振幅平面空气波的传播"一文[1-2],这可能是他在数学物理领域中最伟大的贡献,这个问题属于应用数学(空气动力学)

领域。在他所处的时代，科学家们对空气动力学的认识处于实验认知的阶段。尽管 Airy, Euler, Bernoulli, Stokes, Navier, Poisson, Helmholtz 等人做了一些科学分析，但由于描述流动的控制方程是非线性偏微分方程，在非线性波，特别是冲击波出现的情况下，当时的人们根本不知道如何面对。黎曼选择了"激波管"（shock tube）实验这一具体问题进行研究，不仅开创了空气动力学现代研究的先河，也奠定了现代非线性偏微分方程的基础。

（1）在空气动力学领域，他给出了非线性波（不只是线性声波）的数学描述，特别是冲击波两侧的气体状态之间的关系，这一天才的想法现在看来仍是无法想象的；

（2）在数学领域，首次研究非线性偏微分方程，特别是对间断解（广义解）的构造，在微分方程领域是大大超前的工作。偏微分方程的广义解理论由 Schwartz（1915—2002）在一百年后才完整建立，由此可见黎曼工作的开创性。在专业研究者看来，他还开创了相空间与物理空间统一的特征分析方法。

现在让我们体会一下这位巨匠的选题。他从具体的"激波管"问题入手，而不是贪图空洞的一般性；洞察问题的根本（非线性），而不是关注细枝末节（停留于线性理论）；打破常规的思维，而不是拘泥现有理论不放。激波管装置是力学实验室的最基本装置，用来观察流动的状况。黎曼从这里入手，提炼问题，紧紧抓住非线性"间断"本质现象。很多人可能会说这个问题没有见过，没法做，可是真正"原创性"的工作又有谁预先见过呢？可见不要相信目前指南上所说的所谓"热点问题"。指南上的"热点"也许在工程上有点意义，就像衣服的流行色一样，过季就没意思了。科学上的真正热点是原创的、流芳千古的。

因此推荐读一读黎曼这篇巨作[1]（原文见文[2]）。

二、Lax 具体中的一般性

Peter D. Lax 是当代最伟大的应用数学家之一，生活在一个让他足够自信的时代（虽然作为犹太人后裔从匈牙利移民到美国），他的工作总是轻快且简洁，深刻且广博，读他的文章如听洗涤心灵的钢琴曲。可积系统中的 Lax 对，双曲守恒律中的 Lax 熵条件，科学计算中的 Lax-Friedrichs 格式、Lax-Wendroff 格式、Lax 等价定理、Lax-Wendroff 定理，偏微分方程中的 Lax-Milgram 定理等一连串以他名字命名的定理公式表述简单明了，载入史册。

Lax 的工作在某种程度上是难以模仿的，这或许与他的生活环境有关。Lax 生活在二战后的美国，他的导师是应用数学大师 Friedrichs，工作在应用数学中心 Courant 研究所，周围大师云集，如 von Neumann 等，这些都为他轻快的研究风格奠定了基础，但这不妨碍欣赏他像如来佛一样如何将小猴子们罩在手心。我们选择他 1957 年的名作 "Hyperbolic systems of conservation laws Ⅱ" 进行鉴赏[3]。

二战后，空气动力学的研究被提高到非常高的地位，计算流体力学一直引领和驱动着科学计算和应用的发展，以流体力学可压缩欧拉方程为核心模型的双曲守恒律研究成为重要课题，冲击波理论成为核心中的核心，Courant 研究所作为应用数学中心自然引领着这方面的研究。在这篇杰作中，Lax 从具体欧拉方程的黎曼问题出发，开始研究一般双曲守恒律组的黎曼问题。具体解决这一问题应该不是他追求的全部，通过这一研究，他发展了很多重要的概念，如冲击波传播应满足的 Lax 熵不等式（熵条件），非线性波一般分析方法（特征场的真正非线性或线性退化等）等；几乎把黎曼的工作全部进行提炼、加工，进行了一般化的阐述。时至今日，这些基本的概念和方法仍然指导着后续的工作，包括理论、计算及其应用。从概念上来讲后续的工作还没有本质上的新突破。

从具体的问题入手，发展出基本概念，总结出应用广泛的深刻定理，并以简单明快的方式展现出来，这就是 Lax 风格。

三、Godunov 应用驱动出的奇葩

在计算流体力学界，没有人不知道 Sergei K. Godunov，他创造的 Godunov 方法奠定了现代计算流体力学的基础[4]。1999 年著名刊物 *Journal of*

Computational Physics 组织了一个专栏纪念该论文发表 40 周年，Godunov 详细回顾了 Godunov 方法的构建过程[5]，给人启迪。

20 世纪 40 年代末和 50 年代初，受到第二次世界大战以及美国和苏联紧张关系的影响，航天航空、武器物理等研究得到极大的重视，计算流体力学也被提高到空前的地位，数值算法的设计成为该学科的中心。苏联科研机构的工作和目前国内大多数学术机构的工作方式是类似的，Godunov 被分配学习和跟踪美国同行的工作，完成相关算法的设计[4]。这一过程对 Godunov 来说是痛苦的，在此期间曾受到项目领导的责难，并让他绝望。让我们看看 Godunov 自己是怎么说的[5]："The development of difference schemes for gas dynamics was parallel to the attempts to formulate and adapt the concept of a generalized solution for quasi-linear systems of equations. Usually the hypothesis about possible definitions and properties of generalized solutions preceded the construction of the difference schemes which tried to make use of those properties. At the same time, I was trying to justify the hypothesis. To my greatest frustration those attempts were unsuccessful, and even worse, they often resulted in contradictory examples. Not surprisingly, those examples put me in despair." 这一外行领导内行、行政干预科学的状况今天仍然没有根本改变。但成功总是赋予有准备的人。当他看到 Laudau 和 Lifshitz 关于气体动力学方程组黎曼问题的表述时，立刻意识到应该将其解的结构加入其中，构造了载入史册的 Godunov 方法，并成为他博士论文的一部分，这是在真正应用驱动下结出的奇葩。不过这一超前工作的发表过程并不顺利："However, it was only in 1959 when it was published in the literature. Until that time I had been trying to submit it to several journals, but without any success. For example, *Applied Mathematics and Mechanics* rejected the article as purely mathematical and having nothing to do with mechanics. One mathematical journal (I do not remember which one exactly) rejected it for the exact opposite reason. After that Petrovskii helped me to publish the work in *Mathematicheskii Sbornik* where he was a member of the editorial board."[5] 由此可见，真正原创超前的工作得到立刻的认可是相当困

难的。想做原创的工作需要勇气和胆识，而跟踪性的工作可能只是发表的周期短点。

我推荐并叙述 Godunov 方法的创造历程，可以看出即使处在被动选题的环境中，只要不惧困难，最终仍然可以做出非常出色的工作。尤其是在 20世纪 50 年代并没有多少人认识到黎曼问题的价值，Godunov 能够创造性地将其作为数值方法的基石难能可贵。幸运的是，在他的工作发表的当年，Godunov 就被授予"列宁勋章"，我们不得不钦佩苏联科学家的水平和自信。

Godunov 方法以黎曼问题的解为基石，反映了黎曼的这一理论（科学）成果在应用方面的重要价值。

四、Glimm 随机中的确定性

James G. Glimm 的工作完全是另外一个风格，他在纯数学和应用数学领域都做出了杰出贡献，他和 Arthur Jaffe 创立了 Constructive Quantum Field Theory，以他名字命名的 Glimm Algebra 影响深远。Glimm 的另一个重大贡献是建立双曲守恒律组解的存在性理论[6]，由此发展出"随机选取"（Random Choice）科学计算方法。他从熟悉并取得巨大成功的量子物理领域来到与流体力学密切关联的双曲守恒律的陌生领域，足见他作为一个"大家"的魄力。

我们欣赏一下他在双曲守恒律方面的工作。他的选题毫无疑问受到 Courant 学派，特别是 Lax 的影响（坊间传说 Lax 让他思考这一问题）。Courant 学派希望应用研究和科学计算应该有扎实的理论基础，相关应用模型的适定性自然成为数学家们应该回答的问题之一。当时数学家们对这个问题的理解基本停留在黎曼问题上（黎曼本人以及 Lax 1957 年的工作），相关空气动力学方面的工作停留在应用层面，Glimm 经过长时间努力完成了名作[6]。在这一工作中，两个重要的关键点是：（1）非线性波的相互作用估计；（2）通过随机选取，巧妙地应用黎曼问题的解作为基石构造逼近解。波的相互作用在应用中已有具体的研究，但 Glimm 天才地引入"Glimm functional（泛函）"，对非线性相互作用波的强度变化情况给予定量描述，从而该研究有了扎实的抓手；而随机选取的引入更为天才，他把物理中的随机性和数学中的确定性（正则性）完美地融为一体，令后人高山仰止。虽然衍生出的随机选取方法产生了很大的影响，但"随机中的确定性"留给人们无尽的思考。

Glimm 这一工作当然源自他对自己"天才"的自信和解决科学难题的勇气。他的工作深邃，他的风格持久。这些在他后来的工作中一直得到体现。

五、张同严谨中的浪漫

张同先生是我的博士导师，一生从事黎曼问题研究。作为学生，我对他了解比较多，这里谈谈他的"黎曼人生"。

他的研究工作主要受到著名数学家 Gelfand 的著作（中译本）影响[7]，那时中国一切向苏联学习，苏联也确实引领众多科学领域的发展，如著名的 Laudau 和 Lifshitz 的系列巨著就是一例；同时由于核武器研究的需要，国内也在研读 Courant 和 Friedrichs 的经典名著"Supersonic Flow and Shock Waves"[8]，因此在当时环境下，选择这个问题研究就是自然而然的了。他也告诉我看到 Gelfand 在文中对双曲守恒律的描述，坚定了自己从事这方面研究的决心。

从 20 世纪 60 年代到 70 年代，张先生和同事们主要从事具有一个空间变量的黎曼问题研究，取得了一系列成果[9]。学生们都知道，他把空气动力学中波的相互作用进行了整理并提炼出一系列结果，如通过对波的相互作用分析研究具有"强波"相互作用的流场性态，非凸状态方程流场中"冲击波"的熵条件等。这些工作结合应用需求，系统且深刻。这段时间张先生的工作"小心且严谨"。

张先生大胆且浪漫的工作应该是"气体动力学方程组的两维黎曼问题"[10]。这一工作不是从一个空间变量向两个空间变量推广那么简单。众所周知，即使线性情形下，一维波动方程和二维波动方程都表现出完全不同的性态，何况非线性情形。在对一维问题进行深刻研究之后，二维问题的研究是不得不面对的，但如何有效提出并构想出"合适"的数学问题是一个巨大的挑战，这时候张先生展现了他"美声"歌唱家的天赋，构想了气体动力学方程组两维黎曼问题的提法，边分析边猜想了其解的结构。

张先生和郑玉玺的"猜想"是不拘一格的。正是由于和"正统"的研究方式不同，浪漫的情怀受到许多外行装模作样的评价不足为奇。这一问题的构思方式可能是最简洁的，包含了可能所有多维非线性波的本质相互作用，如冲击波发射中的"正规反射"、"马赫反射"、涡团的相互作用、冲击波和涡团的相互作用、气体扩散等[11]。即使现在猜想得到严格证明的部分仍然很少[12,13]，但该猜想已成为数值模拟和应用研究的基准[14]。

后记：就像欣赏文学作品一样，需要读名家、品名家。至于如何做学问，可能既要"学"又要"问"。所谓师傅领进门，修行在个人。只要我们热爱并坚持，一定可以"作出"优秀的产品，使之成为"作品"，让自己慢慢品味。

此文是应李常品教授盛邀而作，为年轻学生或科研工作者谈一谈科研论文的写作。这一命题作文，对我来说是一份"很困难"的任务：第一，我的科研论文就是"随性"而发，不像现在的任务那么明确；第二，所谓的经验对不同的人有不同的体会，我的经验不见得适合别人。不过朋友吩咐的事情总要完成，于是我根据自己的学习经历，以流体力学（特别是空气动力学）中的"黎曼问题"为例，体会几位"大家"是如何选题和工作的，以抛砖引玉。

参考文献

［1］Riemann B. 黎曼全集［M］. 李培廉，译. 北京：高等教育出版社，2016.

［2］Riemann B. Uber die forpflanzung ebener luftwellen von endlicher schwing ungsweite［J］. Abhandl Koenig Gesell Wiss，Goettingen，1860，8，43.

［3］Lax P D. Hyperbolic systems of conservation laws II［J］. Comm Pure Appl Math，1957，10：537－566.

［4］Godunov S K. A difference method for numerical calculation of discontinuous solutions of the equations of hydrodynamics［J］. Mat

Sbornik, 1959, 47: 271 – 306.

[5] Godunov S K. Reminiscences about difference schemes [J]. Translated from the Russian by Konstantin Kabin and Natasha Flyer. J Comput Phys, 1999, 153: 6 – 25.

[6] Glimm J. Solutions in the large for nonlinear hyperbolic systems of equations [J]. Comm Pure Appl Math, 1965, 18: 697 – 715.

[7] Gelfand I M. Some problem in the theory of quasi-linear equations [J]. Uspehi Mat Nauk, 1959, 14: 87 – 158.

[8] Courant R, Friedrichs K O. Supersonic Flow and Shock Waves [M]. Interscience Publishers, Inc., New York, 1948.

[9] Chang T, Hsiao L. The Riemann problem and interaction of waves in gas dynamics [M] //Pitman Monographs and Surveys in Pure and Applied Mathematics, 41. Longman Scientific & Technical, Harlow, 1989.

[10] Zhang T, Zheng Y. Conjecture on the structure of solutions of the Riemann problem for two-dimensional gas dynamics systems [J]. SIAM J Math Anal, 1990, 21 (3): 593 – 630.

[11] Glimm J, Ji X, Li J, Li X, Zhang P, Zhang T, Zheng Y. Transonic shock formation in a rarefaction Riemann problem for the 2D compressible Euler equations [J]. SIAM J Appl Math, 2008, 69: 720 – 742.

[12] Li J. On the two-dimensional gas expansion for compressible Euler equations [J]. SIAM J Appl Math, 2001/2002, 62: 831 – 852.

[13] Li J, Zheng Y. Interaction of rarefaction waves of the two-dimensional self-similar Euler equations [J]. Arch Rat Mech Anal, 2009, 193: 623 – 657.

[14] Lax P, Liu X D. Solution of two-dimensional Riemann problems of gas dynamics by positive schemes [J]. SIAM J Sci Comput, 1998, 19: 319 – 340.

数学专业研究生学位论文的写作

姚锋平（上海大学）

华罗庚先生曾经形象地拿"兔子"比喻导师和研究生的关系：导师主要负责给研究生指出兔子在哪里，并指导研究生学会打兔子的本领。研究生则需要从导师那里了解到兔子的位置、大小、肥瘦，并采用从导师那里学到的打兔子本领擒获一只兔子（就是做完论文）。后来有人由此衍生出本科生、硕士生、博士生之区别的"兔子理论"：学习捡"死"兔子（本科生主要学的是那些经过反复验证的固定和稳定的知识，故称为"死"兔子）的是本科生；通过从导师那里学来的方法，学习打一只尚在运动中的"活"兔子的是硕士研究生；在归纳总结的基础上研究未知领域，学习在一片丛林里打一只看不到（导师可以提前确认一片丛林里"是否有兔子"）的"活"兔子的是博士研究生。而本科、硕士、博士的学位论文是区别兔子是否是"死"兔子、还是运动中的"活"兔子以及是否是自己通过不断探索与追捕而逮到的"活"兔子的基础，也是三者学位区别的最本质的体现。博士、硕士研究生学位论文质量不仅关乎研究生的培养，同时也关乎社会主义合格建设者和可靠接班人的培养。另外，研究生的学位论文不仅是判别学位申请者学术水平以及是否授予其学位的主要评判依据，同时也是科研领域的重要文献资料，所以学位论文的撰写是非常重要的。很多学者也已经研究了研究生学位论文的相关写作问题，参见文献[1-5]。这里我们将简单阐述一下如何撰写一篇研究生学位论文。

一、选题

一篇好的学位论文需要选题得当并具有很强的专业性、应用性和创新性，只有选题恰当，才能做到目标清晰、方向正确、道路明确，才能写出较高水平的学位论文。对于数学专业的研究生而言，读研究生的第一年可能需要类似本科生一样学习一些本专业相关的基础知识，等对相关专业有了深入了解之后就应该思考自己的学位论文应该如何选题。这个需要研究生与导师的相互配合，当然更需要研究生自己的主观能动性，需要提前预判你在的那片"丛林"里是否有"兔子"。当然，知道有兔子，逮不逮得到可能就是能力问题，这就需要我们扎扎实实地提前学好各种追踪兔子、捕捉兔子的技巧与方法（专业基础知识与本专业的研究技术、研究方法等）。如果是博士研究生，将来可能需要自己不断地寻找兔子与捕捉兔子，所以在研究生阶段学好如何选题，如何预判在一片"丛林"里是否有"兔子"，并具备捕捉运动中的"活"兔子的本领就显得尤其重要。所以如何选题实际上不仅关乎研究生能否顺利毕业，同时也是关乎研究生未来发展的重大问题。

二、绪论

论文内容安排不当，缺乏合理性、逻辑性和一致性是一些学位论文的普遍问题。这就需要我们在写论文之前首先构思论文的整体框架，而首要的工作就是写好论文的绪论部分。可能有的研究生在写小论文时已经尝试写过一部分的研究背景与研究概况，但是学位论文对于绪论部分的要求往往比以前更高。这就需要我们在撰写学位论文之前大量并广泛地阅读（精读与泛读）一些论文与书籍，从而对课题的研究领域有更系统、更完整、更全面的了解。当然，如果可以找到本领域学术大咖写的相关研究背景与研究进展的综述论文那就可以事半功倍。事实上，绪论部分的撰写特别考验一个研究生的写作能力，也最能体现学位论文的质量高低，所以研究生必须认真地对待绪论部分。通过写好绪论部分可以很好地展现学位论文的创新性或应用价值，同时也可以有效地避免文献掌握不到位而产生的误会。

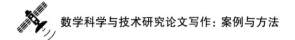

三、正文：主要结论与证明

对于数学专业的研究生而言，因为学位论文主要结论往往是自己已经发表的小论文的内容，所以正文部分在写学位论文之前就可能已经基本确定，由于主要结论与证明细节往往已经经过验证，从论证的正确性来说可能问题并不大。但是很多学生经常把已经发表的英文论文直接翻译成中文，导致阅读起来很别扭，所以这块内容也需要学生好好撰写并反复推敲，从而很好地展现学位论文结构的完整性、逻辑性、创新性、学术性与思想性。

四、参考文献

参考文献部分看起来无足轻重，其实不然。很多研究生的学位论文的参考文献部分经常出现很多规范性的问题，主要问题包括：刊名、期号、页码格式不统一，中英文文献、书籍的格式不一致，杂志名的名称与缩写不正确，更有甚者把作者的名字都写错，这个可能会极大地影响学位论文质量。这部分内容相对简单，但是更能体现研究生对学位论文的重视程度，所以也直接影响学位论文的质量。

五、总结与展望

授人以鱼不如授人以渔，这块内容我们也要认真对待，因为这部分内容可以反映该学位论文作者将来在丛林里抓兔子的能力，同时也能反映该研究生的科研潜力与未来的学术发展。

研究生教育质量直接影响着国家高层次人才培养和创新创造能力的培养，现在我们国家很多"卡脖子技术"需要新一代的研究生去攻克。另外，中国已经成为世界研究生教育大国，研究生教育是我们国家发展、社会进步、应对全球人才竞争的重要基石。2020 年 9 月，《教育部 国家发展改革委 财政部关于加快新时代研究生教育改革发展的意见》明确提出"把论文写作指导课程作为必修课"。研究生应该在读研期间通过自己的坚持不懈、刻苦钻研、严格的学术训练最终在毕业前撰写出一篇优秀的、高质量的学位论文。

参考文献

［1］苏婧. 从混乱到秩序：学术性写作的文献能力［J］. 新闻与写作，2021 (6)：104-108.

［2］岳云强，申晗. 当前硕士学位论文写作中存在的主要问题及对策［J］. 新乡学院学报，2021，38(5)：71-73.

［3］蒋英州. 研究生学位论文质量提升方法探讨［J］. 西华师范大学学报 (哲学社会科学版)，2021(3)：98-103.

［4］王晶，甘阳，杜春雨. 工科博士学术论文发表对学位论文质量的影响 ［J］. 黑龙江教育（高教研究与评估），2019(12)：77-79.

［5］郑淑明. MTI毕业论文写作的探索与实践——以哈工大校训"规格严格，功夫到家"为指导思想［J］. 翻译研究与教学，2020(2)：55-60.

数学论文写作技巧

——针对研究生如何写好一篇专业论文有感

高楠（上海大学）

一、确定研究主题

首先通过导师指导，给出研究课题；或者通过参加学术会议，仔细聆听报告，然后找出别人在思考什么，选定一个自己最有兴趣的研究领域。确定好感兴趣的研究方向，找到相关方向的大量论文，进行泛读。通过泛读，可以对该研究方向的发展背景有一定的认识，同时了解到不同的学者在该研究方向开展了什么样的研究。在此基础上，选出几篇最吸引你的论文进行精读，精读的同时，这些论文的写作技巧和论文结构以及作者的思考方式和证明思路，都可以学以致用。此外，可以与同方向的人交流，在互联网上查找相关信息，慢慢地找出自己感兴趣的问题，然后去解决它。对于自己的结果，可以和导师讨论，确信正确无误后，那么就可以写出来了。

二、撰写论文

书面和口头的数学交流是一个滤波器，透过它可以审视你的数学工作。如果你准备分享你所做的数学之美，那么仅仅写是远远不够的，关键是要努力写好。除了向读者说明你的工作的真实性，当你在写你自己的研究时，你还希望读者欣赏你所做的数学之美并理解它的重要性。

论文一般包括几个标准部分：标题、摘要、关键词、背景、正文和参考文献。摘要和关键词可以让读者对你的论文的主要内容、结构和相关性有一定了解。这一部分是文章的灵魂，至关重要；背景是你论文所做工作的研究现状，给出此方面工作的相关研究；正文部分包括了主要的研究工作，所以对论文结构和排序要仔细揣摩，将工作一步步展开，引人入胜。不要期待一稿定乾坤，写论文最重要的是第一稿，因为从无到有是最艰难的。完成初稿后，再对内容进行检查和更正，尤其是语言尽量化繁就简，让人读着不累。论文的结构，要有一定的逻辑性。

比如写论文背景，可以先搜索出所做工作的相关论文，将这些论文的背景部分浏览一遍，这样对此工作的研究现状了解于心后，便可以开始着手写文章的背景部分。先指明是由哪位学者提出的，基于解决什么问题而提出；之后又有哪些学者做了相关研究，得出了什么结论；最后可以写自己要做的工作，以及文章的整体框架。例如：

Recently, Ringel and Zhang [RZ] investigated the algebra $A = k[\epsilon]$ over a field k of dual numbers (the factor algebra of the polynomial ring $k[T]$ in one variable T modulo the ideal generated by T^2), and thus to the Λ-modules, where $\Lambda = kQ[\epsilon] = kQ[T]/\langle T^2 \rangle$. Moreover, they characterized the perfect differential kQ-modules. Later, Wei [W] showed a close relation with Gorenstein projective modules. Xu, Yang and Yao [XYY] generalized them by introducing n-th differential modules and showed a corresponding result as mentioned above in this setting.

再以写正文内容某一节为例。为了让读者对这一节内容有清晰的了解，首先概述本节的主要思想和主要结论，然后解释本节要用到的数学记号，下面就可以一一展开主要结果了。如果你的主要结果集中于一个定理，那么在主定理前需要一些定义和引理作为准备工作，对于本节用到的定义需要标明参考文献，引理是为主定理的证明做铺垫。证明过程语言的阐述要简单明了，数学公式的排版要整齐，比如长公式可以另起一行居中显示。我们把结果在 LATEX 上敲出来后，再重新看一遍自己的内容，主要看论文中引理和定理的证明思路的正确性；第三次审视本节内容，对论文语言描述再逐字逐句修改一番，同时排除一些语法错误、单词拼写错误等；第四次看本节内容，可以整体看本节内容的结构，以及公式排版。如此反复，论文初稿基本就可定下来了。例如：

（1）背景扩充

Ringel and Zhang [RZ] investigated the algebra $A = k[\epsilon]$ over a field k of dual numbers (the factor algebra of the polynomial ring $k[T]$ in one variable T modulo the ideal generated by T^2), and thus to the Λ-modules, where $\Lambda = kQ[\epsilon] = kQ[T]/\langle T^2 \rangle$. Moreover, the perfect differential kQ-modules are characterized.

修改前：背景现状少

Recently, Ringel and Zhang [RZ] investigated the algebra $A = k[\epsilon]$ over a field k of dual numbers (the factor algebra of the polynomial ring $k[T]$ in one variable T modulo the ideal generated by T^2), and thus to the Λ-modules, where $\Lambda = kQ[\epsilon] = kQ[T]/\langle T^2 \rangle$. Moreover, they characterized the perfect differential kQ-modules. Later, Wei [W] showed a close relation with Gorenstein projective modules. Xu, Yang and Yao [XYY] generalized them by introducing n-th differential modules and showed a corresponding result as mentioned above in this setting.

修改后：增加了相关研究

（2）交换图

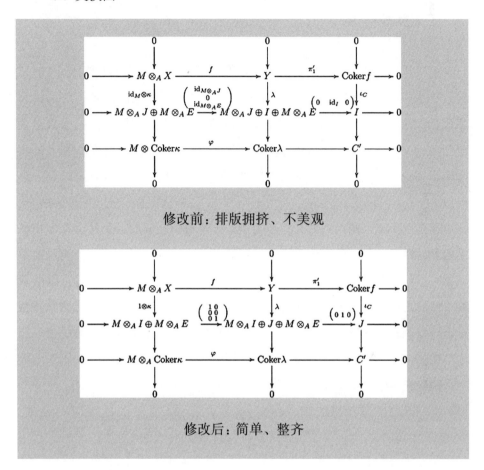

修改前：排版拥挤、不美观

修改后：简单、整齐

（3）正合列

$$0 \to (X, Y, f, g) \xrightarrow{\big(\kappa, \lambda\big)} (L, K, a, b) \longrightarrow (\mathrm{Coker}\kappa, \mathrm{Coker}\lambda, \varphi, \psi) \to 0.$$

修改前：箭头长短不一

$$0 \longrightarrow (X, Y, f, g) \xrightarrow{\big(\kappa, \lambda\big)} (L, K, a, b) \longrightarrow (\mathrm{Coker}\kappa, \mathrm{Coker}\lambda, \varphi, \psi) \longrightarrow 0.$$

修改后：箭头长短一致

三、反复修改

论文整体结构和内容确定好，无需再修改后，这时可针对论文的语言、排版、参考文献格式等进行修改，呈现出一篇漂亮的文章。俗话说百炼成钢，论文也一样，往往最终稿跟第一稿相比总是面目全非的。自己读自己的论文很乏味，并且不易找到错误，再修改的时候，可以从别人的角度来审视论文。为了论文小的层次的提升，要付出大量的精力去一点一点修改。

还是以步骤二中提到的某一节为例，本节初稿完成后，仍然会存在一些不明显的错误。我一般会先将本节内容通读一遍，读的过程中注意有没有单词拼写错误和语法错误，主要是对文字内容进行修改；通读三遍，以防遗漏。然后再通篇检查一下数学公式的标号是否正确，以及数学公式格式排版是否需要调整，观感上尽量整齐。这样反复修改后，便呈现出文章的最后版本。例如：

（1）符号有误

Throughout, let A and B be Artin algebras, $_BM_A$ and $_AN_B$ bimodules with

$$M \otimes_A N = 0 = N \otimes_B M$$

such that $\triangle = \begin{bmatrix} A & N \\ M & B \end{bmatrix}$ is an Artin algebra, where the addition is obvious, and the multiplication is given by

$$\begin{bmatrix} a & n \\ m & b \end{bmatrix} \begin{bmatrix} a' & n' \\ m' & b' \end{bmatrix} = \begin{bmatrix} aa' & an' + nb' \\ ma' + bm' & bb' \end{bmatrix}.$$

This is a special case of the general version of Morita rings in the sense of H. Bass [3]. See also [6, 9, 10, 7].

修改后的正确符号(△→Δ)：

Throughout, let A and B be Artin algebras, $_BM_A$ and $_AN_B$ bimodules with

$$M \otimes_A N = 0 = N \otimes_B M$$

such that $\Delta = \begin{bmatrix} A & N \\ M & B \end{bmatrix}$ is an Artin algebra, where the addition is obvious, and the multiplication is given by

$$\begin{bmatrix} a & n \\ m & b \end{bmatrix} \begin{bmatrix} a' & n' \\ m' & b' \end{bmatrix} = \begin{bmatrix} aa' & an'+nb' \\ ma'+bm' & bb' \end{bmatrix}.$$

This is a special case of the general version of Morita rings in the sense of H. Bass [3]. See also [6, 9, 10, 7].

（2）多括号

$$0 \longrightarrow (0, M \otimes_A DA, 0, 0) \xrightarrow{(0,\,1)} (D(A), M \otimes_A DA, 1, 0) \xrightarrow{(1,\,0)}$$
$$(DA, \mathrm{Hom}_A, (N, DA), 0, e_{DA}) \xrightarrow{(0,\,1)} (0, \mathrm{Hom}_A(N, DA), 0,$$
$$0) \longrightarrow 0 \tag{2}$$

正确格式：

$$0 \longrightarrow (0, M \otimes_A DA, 0, 0) \xrightarrow{(0,\,1)} (DA, M \otimes_A DA, 1, 0) \xrightarrow{(1,\,0)} (DA,$$
$$\mathrm{Hom}_A, (N, DA), 0, e_{DA}) \xrightarrow{(0,\,1)} (0, \mathrm{Hom}_A(N, DA), 0, 0) \longrightarrow 0 \tag{2}$$

（3）多空格

$$\varphi: V \otimes_B M \longrightarrow U \text{ is a monic right } A\text{-map}$$
$$\psi: U \otimes_A N \longrightarrow V \text{ is a monic right } B\text{-map}.$$

删除空格后：

$$\varphi: V \otimes_B M \longrightarrow U \text{ is a monic right } A\text{-map}$$

$$\psi: U \otimes_A N \longrightarrow V \text{ is a monic right } B\text{-map}.$$

四、投稿合适期刊

根据自己的研究方向，选中目标期刊，进行投稿。

五、催稿（≥ 6 个月）

对于学术期刊要有足够的耐心，如果在六个月过去后，仍未收到审稿人的意见，可以写一封询问审稿状态的邮件给接收你论文的编辑，礼貌地询问稿件的录用状况。

六、回复审稿人意见

回复审稿意见一般是要求"point to point"的回复，必须一一进行回答，个人建议是所有问题最好都按要求去做，重视所有的建议，认认真真地修改。对于审稿人提出的每一条建议，都要对照自己的论文进行修改，并在回复审稿人意见时详细标明在文章哪一页哪个位置进行了修改，以及修改了哪些内容。如果你不同意某条建议，用详细的理由加以回复，有可能的话附上参考文献增加可信度。如图所示：

The response letter to the referees' report

We are very grateful to your professional and careful review on our manuscript "Silting objects over the stable monomorphism category of higher differential objects". Your comments are very helpful. We have carefully revised the manuscript based on these comments item by item, and also have reorganized the structure of the manuscript.

(1) The proof of theorem 4.4(ii) is not correct. Please check Step 2 in Page 7.

Author's Response:

(2) In the proof of Corollary 5.3, the two algebras in line 5 are isomorphic.

Author's Response:

七、校稿

论文被接受后，发表之前要进行校稿。对文章仔细检查三四遍，准确无误提交最终版本。

八、纠错

如果在校稿过程中发现错误，就要进行更正，完善文章最后版本。

审稿意见回复写作要点

余长君（上海大学）

二战后，科技进步引发论文数量和期刊种类的激增，20世纪中叶以来，同行评议逐渐成为科技期刊出版的基石。高质量期刊论文的发表都需要经过同行评议这一流程来对论文质量进行评估，只有通过同行专家的认可才能被期刊所接收，回复审稿意见是同行评议中的一个重要环节，如何针对审稿意见作出适当的回复直接影响到论文是否被接受。本文将针对审稿意见的回复给出一些指导性意见，通过一些实际案例，引导读者高质量地回复审稿意见，提升论文接受概率。

一、回复审稿意见的一般原则

1. 认识审稿对提高论文质量的积极作用

科技论文投稿后，一般会由刊物主编/领域编委邀请同领域内的学者评议论文的学术和文字质量，提出匿名意见和判定，主编按评议的结果决定是否适合在本刊发表。这些评议人的评议结果，将直接影响到主编对被评议的文章做出是否录用的评定。一般说来，除非多数审稿人对论文质量提出较为尖锐且难以通过修改而改善的批评，从而导致论文被直接拒稿，否则，论文作者均有机会对编辑/审稿人所提出的审稿意见进行答复，并对论文做适当的修改。这个修改的过程，是提高论文质量的一个重要环节。

2. 尊重审稿人的意见

评议专家大多数时候应该是论文研究领域的小同行，但也不能排除是相对的大同行。因此，论文的评阅意见有些时候能够一针见血指出论文的核心工作，有时也仅仅给出格式、结构、语言等方面的意见，甚至有些时候评议专家的意见会出现一些错误。然而，作为一名研究人员，我们应了解审阅投稿论文是一项义务工作，无论评阅观点高低，意见中肯与否，投稿人均应该保持尊重的态度来看待。

3. 不卑不亢地回复意见

正如前面所提到的，评审意见可能中肯也可能尖锐，对待不同的意见，总的来说应保持不卑不亢的态度，有礼有节地回复每一条评审意见。有所坚持，也有所修正。对于一些"不懂"的评审专家，也应该尽可能地解释，以获得评审专家的尊重。

二、审稿意见回复思路

1. 技术类问题

对于数学学术论文，技术类问题主要包括审稿人对于论文中的概念、理论、模型、方法、数值实验等技术方面所提出的具体意见。这类问题的提出，往往意味着审稿人对论文的研究内容有着较为深刻的了解，往往能够切实帮助论文作者提升论文质量。作者一般需要按照审稿意见，在论文中补充、改进、完善相关研究内容，对于在短期内无法完成的工作，需说明其中具体的原因，但是同时作者应表示出为此做出了什么样的措施来弥补，或者说作者的理由足够充分。

意见示例 1. Numerical method for delayed optimal control has been studied recently by J. T. Betts, S. L. Campbell, and K. C. Thompson. What are the advantages of the proposed numerical method comparing to those recent results, for example, those in: [1, 2, 3].

意见解析：审稿人要求作者解释投稿论文中所提出的用于求解时滞最优控制问题的数值方法与现有方法相比有何优势，这是一类较为常见的问题。一般回复模式为，将审稿人所提出的不同文献中的方法与投稿论文的方法进行多角度比较，若有条件，可以补充一些数值实验。

意见答复 1. In [1], the optimal control problem under consideration is solved by direct transcription, where the system dynamics and the cost function are discretized, giving rise to an optimization problem. This optimization problem is solved sequentially on a sequence of successively refined grids. To be more specific, the system dynamics over the time horizon are discretized to yield a set of equality constraints. Furthermore, both the state and control at the discretized points of the time horizon are regarded as decision variables. For this set of equality constraints to be an accurate approximation of the differential equations, the discretization of the system dynamics must be fine enough. Hence, this will result in a large number of equality constraints. As both the state and control at the discretized points are regarded as decision variables, the number of decision variables will also be large.

If the system dynamics are linear, the resulting approximate equality constraints will be linear, and hence they can be easily solved even with a very fine discretization. However, if the system dynamics are nonlinear, resulting approximate equality constraints will be nonlinear. Thus the corresponding nonlinear optimization problem will involve many equality constraints and decision variables. In [1], it is proposed that the nonlinear optimization problem is solved sequentially on a sequence of successively refined grids. However, the corresponding nonlinear optimization problem will eventually become a large scale nonlinear optimization problem, involving many equality constraints and decision variables, when the discretization becomes very fine. It is computationally expensive for solving such a large scale nonlinear optimization problem, especially with many nonlinear equality constraints. From the examples considered in [1], we see that the system dynamics of the first example are linear and the time horizon is taken as 50, which is quite long. Due to the linearity of the system dynamics, a large number of discretization points are taken. On the other hand, the system dynamics of the second example are nonlinear. For this example, we can see that the time horizon of this optimal control problem is short and the number of discretization points is taken to be small.

In reality, many real world optimal control problems involve also continuous inequality constraints on the state. These continuous inequality constraints will also need to be discretized to yield a set of nonlinear inequality constraints. However, there is no guarantee that these continuous state inequality constraints will be satisfied in between the discretized points, unless the number of discretization points is very large.

For the approach proposed in this paper, known as the control parameterization approach, we only carry out the parameterization of the control function as a piecewise constant or piecewise linear function. The system of differential equations corresponding to each given control is solved by using the 6th order Runge-Kutta differential equation solver. The accuracy of the solution obtained is high. In this approach, the optimal control problem is also approximated as a nonlinear optimization problem, where the cost function is to be minimized subject to the constraints on the state and control. However, the differential equations are not approximated as equality constraints; they are solved, for each given control, by using the 6th order Runge-Kutta differential equation solver. The solution of the differential equations obtained is then used to define the cost function and the constraint functions. The nonlinear optimization problem is of much lower dimension in terms of the number of decision variables and the number of constraints. In particular, there are no nonlinear equality constraints resulting from the discretization of the system dynamic. As the solution of the system of differential equations is solved by using the 6th order Runge-Kutta differential equation solver, the solution obtained for each given control is expected to be very accurate. However, it tends to be time consuming if the differential equations are stiff.

The control parameterization approach has been used in conjunction with the time scaling transformation (see [4]) for solving many classes of optimal control problems. However, the time scaling transform has problem when it is applied to optimal control problems, where there are time delays appearing in the state and/or control of the differential equations. The contribution of this paper is to propose a hybrid time scaling transform, which can be used in conjunction with the control parameterization technique for solving the time delayed optimal control problem considered in this paper. We expect that our approach will produce superior solutions when compared with those obtained by direct transcription. To verify this, we apply our method to solve the two examples considered in [1]. From the numerical results obtained, we can see that the solutions obtained by our method give rise to lower cost values.

For [2], its focus is on an optimal control problem, where its system is governed by partial differential equations with time delays. For the optimal control problem considered in this paper, the system is governed by ordinary differential equations with time delays. For the optimal control problem considered in [3], its time delays appear only in the control. For the method developed in this paper, it is applicable to optimal control problems with time delays appearing not only in the control but also in the state.

意见示例 2. The performance of the proposed numerical algorithm is demonstrated on two examples from [5], which was published more than twenty years ago. Given the advances in the computational optimal control field, it would be better if the algorithm is compared with more recent numerical methods.

意见解析：审稿人认为作者进行数值实验所使用的算例较为陈旧，要求作者进行更多的数值实验，并且与更多方法进行比较。这里应该按照审稿人要求去做更多数值实验并进行比较.

意见答复 2. We have removed the first example, and added three new examples from [1, 6] to demonstrate the effectiveness of the new method.

意见示例 3. When the authors use their function μ to get equation (18) they claim that "... the fixed time delay h becomes variable in the new time horizons", even though the only way h appears in equation (18) is in the form $\mu(s \mid \theta) - h$. I do not understand what the authors mean by saying that h, which is a constant, becomes "variable" after one uses μ instead of the original time t.

意见解析：审稿人对某一技术细节有部分质疑，这里应该详细解释，使审稿人明白。

意见答复 3. The time scaling transform is a useful technique which maps the varying time points in the original time horizon into prefixed knots in a new time horizon through the introduction of an additional equation. However, for time delayed system, after the time scaling transformation is applied, the new time delay in the new system on the new time horizon becomes s_{delay}, which is a variable, depending on the optimization parameter vector θ and the time point s. To be more specific, this new time delay s_{delay} satisfies

$$\mu(s_{\text{delay}} \mid \theta) = \mu(s \mid \theta) - h.$$

It appears that the explanation in the original paper is not clear enough, and so this point is explained in further details in our revised paper.

2. 书写类问题

这类问题通常表现为三种情况：（1）审稿人对论文架构提出具体意见；（2）审稿人指出论文总体的英文表达不够好；（3）审稿人对某些文字/记号的使用有意见。对于第一种情况，一般建议按照审稿人的要求对文章架构进行适当修改，以使得审稿人满意；而对于第二类问题，若有条件最好请英文论文润色机构对文章进行整体修改，或者可以求助于英文写作能力更高的人予以指导。至于第三种情况，一般建议直接按照审稿人的意见进行修改，否则应该在回复意见中予以解释说明。

意见示例 4. In Section 3.2, it is suggested to reorganize the introduction of the new time horizon to make it easier to be understood, although this part came from the author's previous work. For example, specifying that the value of $\mu(\gamma \mid \theta)$ is the corresponding time point in the original time horizon for the given time point γ (i.e., $t = \mu(\gamma \mid \theta)$) in the new time horizon would be helpful for the readers to understand this part.

意见解析：审稿人建议对论文引言中所提出的某概念进行更加详细的介绍，并给出了具体的指导意见，应该按照审稿人的要求进行修改，并在后面附上具体的修改内容以使得审稿人较为清楚地看到对其意见的回应。

意见答复 4. We have added a more detailed explanation on timescaling transformation at the beginning of Section 3.2 to make it easier to be understood.

意见示例 5. In Section 3, the relative intermediate time point $\tau_j \in [t_0, t_T]$ are expressed as equation (23) in Definition 3.1. This convex combination should be written as

$$\tau_j = t_0 + \alpha_j (t_T - t_0),\ \alpha_j \in [0, 1].$$

Hence, the expression of τ_j should be corrected in the following part of this paper.

意见解析：审稿人对某个符号的定义有特定意见，若作者认为该意见合理，则直接进行修改即可。

意见答复 5. We have changed the expression for τ_j in the revised version of our paper.

意见示例 6. Please check the following notations are right or not?

(1) In Theorem 1, Line 2, $\gamma \in (-\infty, T)$ or $\gamma \in (-\infty, p)$?

(2) In Section 4, Eq. (16), whether the second term is the partial of $\tilde{H}$ with respect to $\tilde{y}$ but not with respect to y?

意见解析：审稿人对某个记号的正确性有疑问，作者应仔细检查，做出相应的答复。

意见答复 6. (1) In Theorem 1, Line 2, $\gamma \in (-\infty, T]$ has been replaced by $\gamma \in (-\infty, p]$.

(2) In Section 4, Eq. (16), we believe that the second term is still the partial of $\tilde{H}$ with respect to $\mathbf{y}$, but not $\tilde{\mathbf{y}}$.

意见示例 7. (1) Page 1, line 9, in Abstract, "constraint function", may be changed to "constraint functions".

(2) Page 2, line 26, "Section 6 conclude", should be "Section 6 concludes".

(3) Page 3, line 7, something is missing before "is called a feasible control.", maybe u.

意见解析： 审稿人对某些用词做出直接修改，除非出现明显的歧义，否则建议按照意见进行修改，这类意见可以集中在一起，用一条回复即可。

意见答复 7. We have revised our paper as suggested.

3. 概括类问题

这类意见通常没有针对论文中的某个点展开评论，而是笼统地指出论文存在诸如缺乏创新、贡献不清等问题。对这类问题，建议在回复意见中以耐心谦虚的态度进行一一答复。

意见示例 8. What is the main contribution of the manuscript?

意见解析： 这里建议将论文的贡献以更加清晰明了的方式表述出来。

意见答复 8. Control parameterization method used together with time-scaling transformation is an effective approach for solving time-delay optimal control problems. The gradient of the cost and constraint functions are obtained by variational approach, and this approach is relatively inefficient when the discretization for the control function is dense.

The major contribution for this paper is that we proposed a new gradient computational formula for solving optimal control problems with time-delay. Compared with variational approach, the computational efficiency for the gradient of the cost and constraint functions is much higher by using the formula that we proposed. Numerical results show that our method is highly effective.

4. 其他问题

审稿人可能是投稿论文方向的专家，对该领域比较熟悉，审稿人推荐的参考文献，建议一定要阅读和引用。在某些情况下，审稿人对作者的工作存在误解，从而给出一些不太合理的意见。碰到这种情况时，作者一般可不按照审稿人的意见修改论文，但必须在修稿说明中针对误解进行耐心的解释。

三、审稿意见回复的结构

1. Cover Letter——写给期刊编辑的引言段

Cover Letter 1: We wish to thank the reviewers and Associate Editor for carefully reading of our paper and providing valuable comments and suggestions. We have taken these comments and suggestions into careful consideration and the paper has been substantially revised and improved. Our responses to the comments and suggestions made by them are detailed below.

Cover Letter 2: Your comments and those of the reviewers were highly insightful and enabled us to greatly improve the quality of our manuscript. In the following pages are our point-by-point responses to each of the comments of the reviewers as well as your own comments. Revisions in the text are shown using yellow highlight for additions, and strikethrough font for deletions. In accordance with reviewer's suggestion, we have revised .... We hope that the revisions in the manuscript and our accompanying responses will be sufficient to make our manuscript suitable for publication in ....

2. Response to Comments——对审稿意见的逐条回复

点对点地针对审稿意见进行回复，详细说明修改原因，如何修改，在论文里具体修改了哪里（具体到哪一章节哪一行），需说明清楚。多个审稿人的意见要分开回复，不要穿插在一起。修改的地方可以视情况在原文中使用高亮、审阅模式、批注等进行标记。你可以不同意审稿人的意见，但需委婉地阐明理由，有科学的依据作为例证。

3. List of Changes——论文的主要变化

应详细描述对稿件做出的主要改动，例如，对任何图表的删除。向期刊编辑说明如何发现这些改动的地方。例如，应描述"新增部分以黄色背景突出显示"。

示例：We have made an essential revision on the original paper, in particular, Section 2 (Dynamic system with pre-given terminal time) and Section 3 (Dynamic system with undetermined terminal time) in the original paper are now combined into one Section in the revised paper. The surrogate conditions are needed to replace the characteristic time control constraints no matter the terminal time is fixed or unknown. Furthermore, we introduce constraints on the duration of the control region. Two examples are solved to demonstrate the effectiveness of the proposed method.

— On Page 6, we have replaced $t = \mu(s \mid \boldsymbol{\theta}) = \sum_{l=1}^{\lfloor s \rfloor} + \theta_l + \theta_{\lfloor s \rfloor + 1}(s - \lfloor s \rfloor)$, $s \in [0, p]$ with

$$t = \mu(s \mid \boldsymbol{\theta}) = t_0 + \sum_{l=1}^{\lfloor s \rfloor} \theta_l + \theta_{\lfloor s \rfloor + 1}(s - \lfloor s \rfloor), \ s \in [0, p].$$

— On Page 6, we have replaced $\tau_j \in \left[\sum_{i=1}^{\kappa(s_j|\boldsymbol{\theta})} \theta_i, \sum_{i=1}^{\kappa(s_j|\boldsymbol{\theta})+1} \theta_j\right)$ with

$$\tau_j \in \left[t_0 + \sum_{i=1}^{\kappa(s_j|\boldsymbol{\theta})} \theta_i, t_0 + \sum_{i=1}^{\kappa(s_j|\boldsymbol{\theta})+1} \theta_i\right).$$

— On Page 9, we have replaced $\tau_j = \alpha_j(t_T - t_0), j=1, \ldots, m$ with

$$\tau_j = t_0 + \alpha_j(t_T - t_0), j=1, \ldots, m. \tag{1}$$

— On Page 9, we have replaced $\boldsymbol{v}(\alpha_j(t_T - t_0)) = \boldsymbol{\zeta}_j, j=1, \ldots, m$ with

$$\boldsymbol{v}(t_0 + \alpha_j(t_T - t_0)) = \boldsymbol{\zeta}_j, j=1, \ldots, m.$$

— On Page 10, we have revised and proved Theorem 3.1.

Please refer to the concrete revision on Page 5 of the response. Note that the page numbers remarked in red are referred to the original version of our paper.

四、审稿意见回复 TIPs

从收到编辑回信至回复意见，应在规定的时间内返回修改好的稿件及对审稿意见的回复，如果无法在规定的时间内修改及返稿件，应提前与期刊编辑沟通好。一般说来对于 minor revision，建议一周后再回复；而对于 major revision，建议回复时间在四周左右。

How to write a math paper that can be read without tears

Olga Y. Kushel (Shanghai University)

1 Introduction

Recently, I have received an invitation to write a paper entitled "How to write a math paper?" for postgraduates and young scholars. My first reaction was a huge surprise. How could I teach someone to write? Indeed, I have not written a perfect math paper ever in my life. Who would have thought I could teach the others? Well, I have published some papers, even in good journals. But sometimes my style is too heavy, notations are too dense, and my English is not so good. Hence the paper of my dreams is still in future plans. So, what can I tell these young scholars? They have Halmos[3], Higham[4], the Elsevier course of scientific writing by Borja[1, 2], and many others. One may take some pieces of good advice even from the interview of Stephen King[5] (of course, he is not a mathematician but he knows how to write). Obviously, I would never write better than Halmos but I can share my own experience. At least, it would be something new. And I would like to start from the best advice in my life: **write a paper you would like to read yourself.**

Why do we need any paper about writing math? One of my students asked me. Isn't it clear, that the most important thing in mathematics is your idea? Once you have got your results, you are almost done. Unfortunately, not yet. You have to make some efforts to transfer your results to other people, even to your own supervisor and co-workers. Otherwise even the most prominent results will just die inside your computer, unread and forgotten.

2 Wow! You have an idea

Here, I do not cover the topic how to find a mathematical problem to solve. Probably, I would think about it later. Assume you are a PhD student or a postdoc and the problem is provided by your supervisor. You have been spending all days and nights trying to solve it for a long time, or conversely, suddenly have obtained a solution in the very first hour. It does not matter how, but you have got an idea to share with the world. So, my first advise is: please, write it down. Just do it immediately. The ideas, which are not written correctly, can be forgotten very soon. And you may regret a lot of it. Do not worry if you will find a mistake in your first idea and it does not survive until the next morning. Mistakes are just a normal part of our life. Do not be so sad. At least, now you know what **not** to do and which direction **not** to go in your research.

Now assume your idea is totally correct. Indeed, it is so beautiful that you are obsessed with the question: is it new? If so, then it may result in a good paper. But if not, and you are just reinventing the wheel? If your initial problem is listed among unsolved ones, then it will be easy enough to check. The other thing is, if you have a particular case of some very general question that had obtained a lot of attention before. Then you have to do a job, finding out what you have really obtained and how to connect your result to the earlier ones, if it is something new, or just a new way to prove something already known. Hence, the probability that you still have something to write about is very high.

Sometimes you have a crazy feeling that new results on your topic are appearing faster than what you read previously, and soon you will be taken to

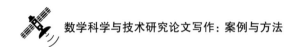

the ocean by a tsunami of information. Relax, this is not true. You need to develop your ability to find key pieces, e. g., starting your reading with the most cited papers.

3 Writing like cooking a soup

Do not over-estimate the difficulties of math writing. Let us consider it as a routine but fruitful process. For example, like cooking a soup.

How do you start a soup?

1) **Write down all the instructions and the ingredient list.** Here, the instructions are all the steps of the proof of your main result and the ingredients are all the definitions and supplementary results you will need at each step. Before starting writing the text, you should state your main results and write down all the proofs as clear as possible. Double-check each step of the proof. You should learn, which technical lemmas you will need and what for. You should think how to arrange them, what is the best way to introduce them to your reader. This leads you to some plan of organization.

2) **Go and find all the ingredients.** Make a run through the literature collecting all the definitions and technical results you will need at each step. Think of the sources you will refer to. There exist a huge number of books and papers and many of them are great in their own way. One topic is perfectly covered here, and for the other one, it is better to see there. A great advantage of a paper is, when a complicated topic is shown in a simple and clear way.

3) **Sauté your ingredients if necessary.** Re-state the existing results in the necessary form using the matching notations, state and prove your own lemmas.

4) **Add your ingredients to your large soup pot following your own menu.** Place all the results into the text following your plan of organizing the text.

5) **Taste and adjust.** Read and re-write.

6) **Keep cooking, adding spices or cream if necessary.** Keep writing, adding introduction, examples, counterexamples, discussion, conclusions.

7) **Taste and adjust again.** Read and re-write again.

Well, you are done.

And some more words about the flavor of your soup. Emphasizing the necessary sides of your results and mentioning their possible applications is like choosing a flavor. Obviously, by choosing the right flavor you can make a delicious soup using the ingredients you have on hand. Hence writing a good introduction which mention the recent results on the topic and showing where they are applicable would help a lot.

A colleague of mine has just said: your text is a pot of water of words with a cup of bouillon of references and slices of meat, and vegetables inside are your own results. So, the stronger the results are, the more tasty your soup would be. However, do not overcook or undercook. Even the best ideas could be spoiled by very bad writing and you soup will go directly to the rubbish box.

4　Key points of math writing

Analyzing a number of writing advices, let us look at them from the following two perspectives. The first is: we are mathematicians and we are lucky enough. No need to dress up the vocabulary. No need to include any poetry, romantics and mystics. Just clear writing using short sentences and simple grammar constructions (see, for example, [4]). Long sophisticated phrases are as unnecessary as decorating your conference presentation with pink unicorns. However, when we think of writing mathematics, we actually think of communicating something that is actually considered difficult to be understood. Hence if mathematics is written by means of symbols and formulas only, it cannot be considered well written, even though it is dedicated to mathematicians only.

The second perspective is that we are mathematicians and we are unlucky. We do live in the real world and we must respect the reality. Therefore the advice to go where the story leads you is definitely not for us. Assume we have a conjecture to prove or disprove. If the question is answered "no" by a counterexample it should be shown from the very beginning, e. g., by entitling the paper "Not all operators of class A are stable". If the answer is affirmative one can write "One more class of stable operators is found".

The paper is not a novel. From this point of view, our writing is boring. We do not have to keep some mysteries until the very last pages. Conversely, we should briefly describe all we have done in the abstract. Moreover, a talented writer can tell the story which is as old as the mountains in a new way, and it will sound like a new story. But in math no fresh idea means no paper. Even to write a good textbook requires a fresh idea how to organize the content and how to write it in the simplest way. This is a pretty hard job.

Let us write down some more points about math papers you should keep in mind.

1) A math paper requires slow reading. It takes your time to read, to follow the proofs, to think of the examples and discussions.

2) It is readable selectively and in pieces. Well, most of the readers never buy a novel to read just one chapter from the middle of it, and then put it on a shelf. But concerning math books—yes, we do like this. Sometimes we need just the definitions and properties of concrete objects. Thus your theorems should be readable and understandable without reading the whole of your paper.

3) It can be used as a reference. A reader should be able to refer the necessary statement easily.

4) It should inspire ideas. As far as I know, no one reads mathematics for pleasure. Every reader needs the results to apply to his own research and

the techniques to follow or to generalize.

5   Writing advice

5.1   Do not write too long

If you write a huge-volume paper, you will take more time and, to be honest will have less publications. It is also more difficult to publish a long (more than 20 pages) text. Many journals have some restrictions for the volume of contributions. Also, it is much more difficult to write a long paper than a short one. Many of us have heard a piece of advice to write using short clear sentences. The same is true in case of expressing your ideas by means of short well-structured papers. Do not try to add to your paper all the results you have obtained and everything you have learned about the subject. Using our cooking analogy, the text is just a soup, not a pizza or Irish stew, where any ingredient is ok. Every text has its own internal logic, and it would not be good to break it. It is much better to write another text after.

Some old mathematicians say, good math writing is like a good poetry — every symbol has its own place. Bad thinking never results in good writing. The crucial question here is "What for?" Any notation, any definition, any lemma should be viewed from this point.

5.2   Read and re-write

Do not start writing any text until you state correctly your main results and clear up all the proofs. Otherwise, when a part of the final section has already been written, you may notice a gap in the proof. If you are lucky enough, it could be avoided by including some additional lemmas or putting some more restrictive conditions. But even in the best case, you would have to re-write a lot. Hence, if you write in details your main proof before starting the text it would save you a lot of time. Otherwise you have to re-write over and over again.

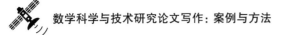

I prefer to start writing from the main part. Often, the core of a short paper is the last but one section, which contains the crucial result. But when you are re-writing, better go from the beginning to the end. "Write in spirals" — advises P. Halmos. Note that every section depends on all the previous ones.

5.3 Mind your notations

Do not introduce new notations for well-known and commonly used objects. Conversely, do not use commonly used notations for something new and different. "The best notation is no notation whenever it is possible" — says Halmos[3]. When the paper is dense with notations, it is not easy to read. You need to think how to avoid this. Do not introduce the notations that resemble each other. To write like

$$\int_P \mathbf{P}(p)\,\mathrm{d}P$$

is certainly not the best idea. Check if the notations are the same through the whole of text. Misprints like $\dot{y} = \dfrac{\mathrm{d}x}{\mathrm{d}t}$ or $\Delta x: \ = t_{i+1} - x_i$ can make your paper totally unclear.

5.4 Simplify your proofs

As a lazy person, I am always looking for a simpler proof. The first reason to follow this advice is, it is so much easier to write; the second is, it is much easier to read (and to publish also). Avoid anything over-complicated. If you do not understand, what for do we need this or that condition, try to prove the theorem without it. To know another proof is always an advantage.

To solve a problem is like to get through the labyrinth, from one point to another. If you have done the proof, you have got one way how to beat the labyrinth. After thinking of the same problem over and over again from different perspectives, you will be able to make a map of the maze. Having a map, you can see all the labyrinth from above and can find the shortest and the most convenient way to get from point A to point B. That is why it may be important to leave you results for some time. The very fresh proof is just one path through nowhere, and we need a complete picture.

Some of us are afraid that with the development of (ATP) (Automated Theorem Proving), there will be no need of math research and math writing any more. However, the proofs generated by automated theorem provers are typically very long, and the problem of finding shorter proofs which are understandable and checkable, is a crucial one.

5.5 Make it more clear

Here, let us consider some tips for clear writing.

1) **Always mention the concrete reference.** No need general phrases. If you are referring to some results, better to find the original paper than use cross-referring.

2) **Give definitions.** If you are involved into the topic, you know much more about what you have written, and sometimes can not notice the gap, i. e. , the absence of some facts that are well-known to you but not to your reader. You should think about your reader and his level of knowledge. Believe me, whoever your reader may be, he will not be Harry Potter. So he could not extract what does that mean directly from your brains. Moreover, if you define something, make sure your definition is the simplest one and the defining class does not coincide with something already known. Don't define "classic French pureé made from potatoes" when "potato mash" is already known and well studied. Stating the known result, you should think, how and where your are going to apply it. To recall the statement, it would be probably better not to use the most general form of it (e. g., Minkowski's integral inequality if you need just the triangle inequality).

3) **Provide clear proofs after statements.** Mark where the proof starts and where it ends. If it is double-sided, first prove one direction then another.

The construction when one-direction proof is interrupted by another direction proof and is continued after it, is less clear.

4) **Do not use too many phrases like "it is obvious".** And also "it is easy to see", "it is not difficult to prove" …

5) **Your paper should be well-structured.**

6) **Mind the level of difficulty.** When your paper includes too much explanations for elementary things and just some general phrases for difficult ones, it is definitely not a good style.

7) **Whenever it is possible, avoid complicated calculations.**

8) **Mind the typos.** Some extra brackets, missed signs or forgotten constants can cause your reader's obsession.

5.6　Write in English from the very beginning

Believe me, it is important. Otherwise you will have to re-write everything and finally, will result in two papers, an English one, and the initial one in your native language. Mind the grammar. Whatever you write, from horror to neurobiology, you need grammar. If you are not sure in some phrasing or grammar construction, better do websearch for it. The more complicated your sentences are, the more problems will be there.

Read in English: any kind of reading in English will enrich your language. The obligatory reading is at least some papers from the journal where you are going to submit your own one. Better choose the new papers, because the writing style is changing. And read the instructions for authors from the journal webpage.

5.7　Include examples

Examples illustrates your theorems. They allows us to see what is going on. By the way, there were statements in matrix theory, that were claimed to be proved (and even published) and then disproved with the help of 2×2 counterexamples. Also note, that putting any additional restrictions, we narrow the class of objects we study. So, introducing any kind of objects make sure they do exist. We do not need any fantastic beasts in mathematics. Imagine, you have written a paper, in which remarkable properties of some Banach spaces are described. And the next year someone else would show that such spaces can not exist. Hence everything should be illustrated.

6   Time management

Finally, how to find time for writing? Some of us are students. Many of us have to teach. All of us enrich our knowledge by reading new papers, attending conferences, participating at seminars and discussions with colleagues. And of course, every person has a personal life outside the university. So, when to write? How to finish your PhD thesis in time? What to do when the deadline is approaching? A very common point is that every single day you should write at least one line. This advice appears again and again in papers and interviews of famous authors — both scientists and writers. I possibly could add one more: if you can not make yourself to seat and write, just promise yourself to work for five minutes. Five minutes, no more, that's so easy. The main trick is, when you finally starts, you will get involved into your writing, and will work for much more than five minutes. It is hard just to get started. But, as Stephen King says in[5], "you have to cut out the unimportant background chatter. That means no Twitter. That means not going to Huffington Post to see what Kim Kardashian is up to." I know, it is difficult, but it is worth doing. So, keep writing, searching for the results you need, reading and again writing. The proper time management matters whether you are writing a novel or a scientific paper.

7   Conclusions

Write if you have something to say to the world. Write if you can not stop writing. Your first paper will not be perfect but you can turn to the perfection like x turns to infinity. The result will be a paper you would love to read yourself.

参考文献

［1］Borja A. Six things to do before writing your manuscript，in series How to prepare a manuscript for international journals［EB/OL］https：//www. elsevier. com/connect/six-things-to-do-before-writing-your-manuscript.

［2］Borja A. 11 steps to structuring a science paper editors will take seriously，in series How to prepare a manuscript for international journals［EB/OL］. https：//www. elsevier. com/connect/11-steps-to-structuring-a-science-paper-editors-will-take-seriously.

［3］Halmos P R. How to write mathematics［J］. L'Enseignement mathématique，1970，16：123 – 152.

［4］Higham N J. Handbook of writing for the mathematical sciences［M］. SIAM，3rd Edition (2019).

［5］King S. 50 Pieces of Stephen King's greatest writing advices［EB/OL］https：//getfreewrite. com/blogs/writing-success/stephen-kings-greatest-writing-advice.